A PHILOSOPHY OF
THE INSECT

A PHILOSOPHY OF THE INSECT

JEAN-MARC DROUIN

TRANSLATED BY

ANNE TRAGER

Columbia University Press   *New York*

Columbia University Press
Publishers Since 1893
New York   Chichester, West Sussex
cup.columbia.edu

Translation copyright © 2019 Columbia University Press

Philosophie de l'insecte by Jean-Marc Drouin copyright © 2014 Editions du Seuil

Library of Congress Cataloging-in-Publication Data
Names: Drouin, Jean-Marc, author.
Title: A philosophy of the insect / Jean-Marc Drouin.
Other titles: Philosophie de l'insecte. English
Description: New York : Columbia University Press, [2019] |
"Philosophie de l'insecte by Jean-Marc Drouin, 2014 Editions du Seuil. |
Includes bibliographical references and index.
Identifiers: LCCN 2019006036 | ISBN 9780231175784 (cloth) |
ISBN 9780231175791 (pbk.) | ISBN 9780231540728 (ebook)
Subjects: LCSH: Insects—Philosophy.
Classification: LCC QL463 .D7513 2019 | DDC 595.7— dc23
LC record available at https://lccn.loc.gov/2019006036

Cover design: Julia Kushnirsky

Columbia University Press gratefully acknowledges
the generous contribution to this book provided by the
Florence Gould Foundation Endowment Fund for French Translation.

CONTENTS

ACKNOWLEDGMENTS

I especially thank Anne-Marie Drouin-Hans, for suggesting the title, but especially for having accompanied me throughout the writing of this essay. For their careful reading of the entire manuscript, I also thank Bernadette Bensaude-Vincent, Colette Bitsch, Frank Egerton, Jean-Jacques Levive, Luc Passera, Annie Petit, and Christine Rollard. Laure Desutter-Grandcolas commented on a communication that prefigured certain aspects of the book. Romain Julliard enlightened me about entomology as a participatory science. Pascal Tassy checked chapter 2 on classifications. Maurice Milgram reread chapter 5, Hélène Perrin chapter 6, and Patrick Blandin clarified my information on the hermit beetle case. This book also echoes a suggestion made by Jack Guichard about the royal metaphor in bees.

A PHILOSOPHY
OF THE INSECT

INTRODUCTION

Foraging bees have an uncertain future. Ravenous grass-hoppers cause devastation. Butterfly wings shimmer. Mosquitoes vector disease. Industrious ants thrive. Enemy wasps attack picnics. Round ladybugs are childlike. Larvae swarm in a half-eaten piece of fruit. Coupling dragonflies form a heart. Praying mantises play out tragic love.

Our representations of insects are many, as are the reactions of fascination or repulsion that they arouse. Scholarly curiosity and the resulting construction of entomological knowledge, far from reducing this multiplicity, reveal its full extent.

The world of insects is marked by twofold alterity. Strange compared to our world, we break theirs into multiple forms. In 1709 the French author Bernard Le Bovier de Fontenelle describes insects as "animals so different from all others, and so different among themselves, that they allow us to understand the limit-less diversity of models nature may have used to make animals for an infinity of other habitats." The sentence is taken from a funeral eulogy written for the physician François Poupart, author of a "History of the Formica Leo" published in the *Mémoires de l'Académie* (Proceedings of the Academy of Sciences) in 1704, and who, according to Fontenelle, had the patience to observe insects and the art of discovering their "hidden life."[1]

The temptation to use superlatives when speaking about insects is great. This can be witnessed in the way Charles Darwin describes bees that manufacture honeycombs and ants with their "slave-making instincts," considered as the "most wonderful of all known instincts."[2] Another example can be found in the 2001 *Classification phylogénétique du vivant* (The tree of life) by Guillaume Lecointre and Hervé Le Guyader. The authors, who are not inclined to speak of prodigies in nature, describe the biodiversity of insects as "prodigious," mentioning that "the number of insect species surpasses the imagination," citing the existence of twenty thousand species of ants.[3] This dizzying number of species is accompanied by a number of individuals even more vertiginous, which sheds light on why it is so hard for insects and people to live together.[4]

Not everyone has shared in this attention given to insects. In 1753 the French naturalist Georges-Louis Leclerc Buffon proclaims in his *Discours sur la nature des animaux* (Discourse on the nature of animals) that "a fly must not hold in the head of a naturalist more room than it holds in nature."[5] This comment is directed implicitly at the scientist and insect aficionado René Antoine Ferchault de Réaumur, as is the barb that one "admires all the more when one observes more and reasons less."[6] What a perfidious and unjust attack. In his *Mémoires pour servir à l'histoire des Insectes* (Memoirs to serve the history of insects), Réaumur proves that one could observe and reason at the same time.[7] He demonstrates that the scientific scope of a study is not measured by the size of its objects but by the relevance of its methods and the pertinence of its questions.

In the following century, Pierre-André Latreille describes a very large number of insect species and attempts to classify them employing the method of natural families botanists had begun to use.[8] At the same time, the French naturalist Jean-Baptiste Lamarck defines invertebrates, distinguishing insects and arachnids, based on their anatomy and physiology.

In everyday language, people commonly use the term *insect* in a broad sense to cover everything from spiders to scorpions. This use of the word *insect* is not only incorrect in terms of vocabulary but is a true misclassification, an error based on lack of knowledge of the biology of these animals. Indeed, paleontology and comparative anatomy confirm that the distinction between insects and arachnids, far from being arbitrary, is justified by their evolutionary history.

Arachnids, however, have a long history of being found alongside insects in books, exhibitions, and articles about these tiny creatures that haunt our homes and our environment. This is far from new. In the 1879 *Souvenirs entomologiques* (Book of insects) by the French naturalist Jean-Henri Fabre, spiders and scorpions occupy a place of honor. Fabre, like Réaumur before him, is more interested in observing behavior than in the labor of classification, although he has perfectly integrated the distinctions Latreille and Lamarck have made. As made explicit in the French title, his memories are deliberately *entomological* and not just about insects.

The relationship contemporary Western culture has to insects cannot be reduced to one of fascination and repulsion. At the most insignificant level, insects provide a wealth of familiar expressions and metaphors, including "being as busy as a bee" or as "crazy as a bedbug," "to have butterflies in your stomach" or "ants in your pants," "to be a fly on the wall" or even a "social gadfly"—someone who is both irritating and stimulating at the same time, a description used by Socrates to describe his method.[9] The most famous rock band of the 1960 bears a name—the

Beatles—one letter off from a group of insects that form the order Coleoptera. Some nursery rhymes and songs mention insects, among which "The Ant" by French resistance member Robert Desnos comes to mind.

Literature and cinema tend to stage insects as a threat. Ants, in particular, claim celebrity status.[10] The 1905 short story by H. G. Wells, "The Empire of the Ants," which was made into a movie in 1977, and the Ants trilogy by the contemporary French novelist Bernard Werber (*Empire of the Ants, The Day of the Ants, The Revolution of the Ants*), which mixes fiction and fact.[11] Another example can be found in the metal-eating ants, imagined by the Italian short-story writer Dino Buzzati, that endanger New York's skyscrapers.[12]

In addition to novels and fictional films in which insects spark anxiety, some photographers and movie producers use close-up images of insects to astonish as much as to instruct, to fascinate more than to worry by showing a "microcosmos" and the "faces of insects."[13] The same inspiration can be found in museum projects, such as the Montreal Insectarium.

In general, insects occupy different places in different cultures. André Siganos's book *Les Mythologies de l'insecte* (Insect mythologies) analyzes the roles insects occupy in the structures of the collective imagination.[14] One of the points of rupture lies in the use of insects as food, common among some people, but excluded by others, even when they consume shrimp, lobster, and other marine crustaceans.[15]

One might be tempted to believe that the ultimate phobia would take the form of humanity fighting off insect invasions of the countryside and cities, but worse has been imagined: a breach of the self. When the hero of Kafka's *Metamorphosis* is transformed into an insect, he loses possession of himself. In *Woman of the Dunes*, a Japanese novel published in 1962 and brought to the screen in 1964, the reader's anxiety stems not from size or invasion but from the hero being trapped like an insect.

Due to overwhelming ecological concerns in the past few decades, the agonizing question "What would become of us in a world overrun by insects?" has been replaced by the question "What would become of us in a world from which insects have disappeared (especially bees)?" This interrogation calls for environmental ethics as well as mobilization of ecological knowledge.

A Philosophy of the Insect is not a philosophy of insects. Rather, it is "of the insect" in the same way it is customary to speak of a philosophy "of the law" or "of art," "of science" or "of nature." It conveys the conviction that the philosopher cannot think about life without including insects, and that he or she cannot take insects into account without questioning entomology and being challenged by it.[16]

Thus philosophy tackles issues as fundamental as size and scale. It uncovers how the concept of insect developed by progressive subtraction of various neighboring groups. Like its literary transposition, ethology focuses on insect behavior and demonstrates to what extent anthropomorphism inhabits the discourse on insects—like a ghost exorcised or an obstacle overcome but never truly dismissed. Philosophy thus challenges the notion of insect societies; it questions the emergence of collective intelligence. By considering the position held by insects in our social and economic life, it meditates on the methodological, cognitive, and practical consequences of disrupting the patterns that distinguish insects as friends or enemies, useful or harmful. By exploring research conducted in fields other than entomology, philosophy discovers a wealth of epistemology. And, finally, the "world of insects" raises the broader question of the "animal world" and that of the possibilities, and limits, of our ethical concerns.

1

TINY GIANTS

W hat are the proportions which will merit your esteem?"[1] By thus questioning those who despise insects, the French historian Jules Michelet emphasizes how, by the standards of common sense, people characterize insects first of all by their physical smallness. Measurement nuances the first impression without contradicting it. A field guide instructs that insects range from "under 1/4 mm to about 30 cm long and from 1/2 mm to about 30 cm across the wings."[2] An Australian phasmid, whose thin body and filiform legs span nearly thirty centimeters, and moths with similar wingspans are exceptions.[3] If we stick to European fauna, the maximum width reached, that of the death's-head hawkmoth, is 12 centimeters, and the length of the stag beetle (*Lucanus cervus*) does not exceed 5 centimeters.[4] So, when reflecting upon insects, it is impossible to escape the point that the largest insects are ten times smaller than we are—and these are extreme and rare cases.

THE FAMILIAR YET COMPLEX
CONCEPT OF SIZE

Size, which plays a decisive role here, demonstrates some astounding characteristics. These become evident when we compare size with two other spatial determinations: position and form.[5] In the absence of some special type of intervention, changing the position of a solid object does not necessarily change the object: my pencil remains the same whether I hold it vertically or lay it down horizontally on the table. On the other hand, when I modify the form of an object I transform it, whether that form refers to its external profile or to its internal structure. Size is more related to the object than position is, but less than shape is. This means it is possible to consider an object as being the same as another that is smaller or larger. This difference in size combined with an equivalent form is found in elementary geometry, where rational beings group together similar triangles and other homothetic figures. It can also be found in tales and myths that our imagination fills with dwarfs and giants. Contemporary literature has not abandoned this approach. The surrealist poet Robert Desnos writes about "an ant that's eighteen meters long." It amuses people as it "pulls along and tows / Penguins and ducks in a carriage load," speaking Latin, French, and Javanese. Smurfs, which were created in Belgium to appeal to a family audience, are little blue elves whose houses are the size of a large mushroom.[6] American cinema has made accidental gigantism of insects the premise for several fantasy films. For example, in Gordon Douglas's 1954 movie *Them!*, authorities call on the expertise of an entomologist to fight giant ants. The title itself expresses how a change in size is enough to make the so-called familiar insect strange and threatening.

Use of the imagination does not necessarily imply the presence of narrative elements. In a fragment of Blaise Pascal's *Pensées,* the philosopher wants us to feel that we are suspended between two infinities. After showing the earth as lost in the universe, he focuses our attention on a mite. Through this tiny animal—today mites are classified as arachnids and not as insects[7]—we are invited to consider "infinites," and, in these, animals, including mites, "in which he [we] will find again all that the first had." And Pascal marvels that "our body, which a little while ago was imperceptible in the universe, itself imperceptible in the bosom of the whole, is now a colossus, a world, or rather a whole, in respect of the nothingness which we cannot reach."[8] If for a moment we forget the extreme terms of *whole* and *nothingness* to focus on progressive reduction, we are struck by the purely fictitious nature of these tiny animals that contain worlds inhabited by replicas of those very animals. Mites, which inhabit flours and cheeses and are considered by the French Littré dictionary to be the smallest animal visible to the naked eye, seem to exist at any scale in Pascal's writing.[9]

Change in size with invariance in form can be found a century later in *Gulliver's Travels.* Jonathan Swift's hero travels to several countries, the first of which, Lilliput, is populated by tiny beings, and the second, Brobdingnag, by giants. Mathematicians who serve the king of Lilliput rate Gulliver's size as twelve times their own. They very rightly deduce that his body has a volume 1,728 times greater than their own and that it is therefore necessary to provide him with food and drink in proportion.[10] Swift does not give such a precise number for giants, but he tells us that when his hero is placed on a table he is thirty feet from the ground. Knowing that our tables have an average height of two and a half feet, we can deduce that the giants are twelve times larger than we are.[11] Gulliver is to the giants what the Lilliputians are to him, which is to say that his size is the geometric mean of the size of the giants and that of the Lilliputians.

Twenty-six years later, other giants appear, with satirical intention, in Voltaire's philosophical novella *Micromégas*.[12] The inhabitant of Sirius is eight leagues high. The companion said that the giant met on Saturn is only six thousand feet tall. Their form and behavior are similar to ours; only their lifespan, which is linked to the size of their stars, is commensurate with their gigantism. In the first chapter we learn that at the end of his childhood, or around the age of 450, the inhabitant of Sirius had been charged with heresy for having written a book on insects, whose diameter was a little less than a hundred feet.

The nature of big and small seems so relative that the same character can successively shrink or grow in size. During her trip to Wonderland, Alice unwisely drank the contents of a bottle and shrank to a mere ten inches tall. One cake makes her bigger, but other cakes make her shrink again.[13] Her creator, logician and mathematician Charles Dodgson, better known by his pseudonym Lewis Carroll, gives a playful flavor to these spectacular changes.

These size-related thought experiments are not only used as a support for a meditation on nothingness and the infinite, and as a pretext for social criticism, they also constitute one of the favorite approaches used in popularizing rhetoric on insects.

In his 1798 *Essai sur l'histoire des Fourmis de la France* (Essay on the history of ants in France), the French zoologist Pierre-André Latreille describes an anthill as a "pyramid, contrasting in its grandeur the smallness of the architect."[14] Thirty years later, the French entomologist Martial Étienne Mulsant, author of an introductory book for the female audience, *Lettres à Julie sur l'entomologie* (Letters to Julie on entomology), echoes the idea put forward by Carl Linnaeus the previous century: if an elephant were proportionally as strong as a stag beetle, it could move rocks and flatten a mountain.[15] In 1858, Michelet judges that beetles, which wear formidable armor and yet move with agility, "reassure us only by their size," and he adds: "Were a man proportionally strong, he might take in his arms the obelisk of Luxor."[16] Closer to home, German evolutionary biologist Bert Hölldobler and his coauthor Edward O. Wilson write, in their quite scholarly popular book *Journey to the Ants*, about an ant nest unearthed in Brazil whose construction "is easily the equivalent in human terms, of the building of the Great Wall of China."[17] The Austrian ethologist Karl von Frisch, famous for his work on bee behavior,[18] which earned him the Nobel Prize for Medicine in 1973, also authored a short introductory book on entomology entitled *Zehn kleine Hausgenossen* (Ten little housemates), published in 1955. He indicated that a flea—*Pulex irritans*—can jump 10 centimeters high and more than 30 centimeters in distance. To give meaning to these figures, he adds that "an adult man wanting to compete with a flea would have to clear the high-jump bar at about 100 metres and his long jump would have to measure about 300 metres."[19] To simplify the analysis, we will limit ourselves to height, but the same reasoning applies to distance. This reasoning is based on the equality of two ratios. The first is that which relates the size of the insect and that of man. This ratio is about one to one thousand. The second is between the performance observed in the insect, a jump of 10 centimeters, and that which is calculated as equivalent for humans, here a fantastic jump of 100 meters. It is this calculated term that at the same time serves as the imagined term.

CHANGING SCALE

Attractive as the imagination may find these comparisons that seem to satisfy reason, those who implement them forget that these size ratios involve changes of scale. In short, if one ignores air resistance, an animal that doubles in size would see its muscle strength (which depends on a section of the muscle and therefore on a surface area) multiplied by four and its weight by eight (since it depends on volume). Similarly, if the size of a flea were multiplied by a thousand, its muscle strength would be multiplied by a million, and its weight would be multiplied by a billion. In other words, if it were larger, the flea would certainly be stronger, but, above all, it would be much heavier. In the end, it is useless to lend our size to fleas or grasshoppers, as it would not allow them to jump any higher.

The same type of explanation applies to the seemingly exceptional force of an ant carrying burdens bigger than itself. As with jumps, we like to imagine the weight we would have to carry to compete with the insect. At first glance, again, the calculation seems quite simple, the weight we would carry would be to the weight carried by the ant what our size is to the size of the ant. This is an illusion that does not take into account the physical consequences of a change in size.

Entomological literature did not always attempt to maintain this illusion, and some authors undertook occasionally difficult demonstrations, even for the general public.

Among the works of the punctilious French entomologist Charles Émile Blanchard, is a popular book titled *Métamorphoses, mœurs et instincts des insectes* (Metamorphoses, manners, and instincts of insects), the second edition of which was published in 1877. After making the usual comparisons to highlight insect performance, he refers to the measurements of muscle strength made by the physician Felix Plateau and then posits that "relatively speaking, the strength of small species is still much greater than that of larger ones," which he explained in one sentence: "body weight increases in proportion to the cube, while strength measured by muscle section only increases in proportion to the square."[20]

We find the same didactic audacity in Belgian author Maurice Maeterlinck's 1930 *The Life of the Ant*. The poet warns us against "the error into which we all instinctively fall when we see ants carrying objects three or four times their own size.[21]" For him, this error comes from the fact that "we do not think of the insect's weight" but only of its length, which is directly perceptible to us. Maeterlinck develops his analysis by referring to an article published in the literary journal *Le Mercure de France* in 1922 and entitled "Remy de Gourmont, J.-H. Fabre, and the Ants." Its author, Victor Cornetz, myrmecologist at Algiers University, refers to a text by Yves Delage that was published in the *Revue scientifique* in July 1913.[22] Supported by the scientific authority of two authors, the second of which was a renowned biologist, Maeterlinck can thus explain to the reader that an ant's weight is proportionally the cube of its size, while its muscular strength depends on the square of its size. This is why, according to Delage, an ant "which can carry a grain of wheat ten times its own weight would be able to carry only a hundredth part of its own weight if it were enlarged to a thousand times its present size."[23] Ants benefit from what physicists call a scale effect.

Although the scale effect is generally ignored by even the knowledgeable public, it has necessarily been known for a long time in technical fields, as it conditions the solidity of constructions. An echo of this empirical knowledge can be read in the passage from *Politics* in which Aristotle asserts that there exists "a determinate size to all cities as well as everything else, whether animals, plants or machines."[24] This applies to the size of a boat that cannot sail if it is too big or too small. In the same way, it applies to a city: "one that is too small has not in itself the power of self-defense, but this is essential to a city,"[25] and if it is too large, it can exist as a people but not as a city with institutions.[26]

John Burdon Sanderson Haldane took up the question of scale without referring to the Greek philosopher in his 1928 essay *On Being the Right Size*. Haldane, a British geneticist, also wrote a series of pop science articles focused on the impact of science on politics. The main idea of this particular essay was that just as every animal has an optimal size, so does every "human institution." Haldane takes the example of the common myth of the imaginary giant flea that "people" think would be able to "jump a thousand feet into the air" and refutes it with the principle that "the height to which an animal can jump is more nearly independent of its size than proportional to it." Evoking the connection between ancient forms of government and the modern democracy, he sees the representative mechanisms as demonstrating the various ways to implement democracy in large states. Next, Haldane tackles the question of socialism, a system to which he is partial, but he restricts himself to only addressing how socialism works in relation to the size of the state. He concludes that the idea that a state the size of the United States or the British Empire could convert completely to socialism was as unrealistic as "an elephant turning somersaults or a hippopotamus jumping a hedge."[27]

ABSOLUTE MAGNITUDE

It is not surprising that Aristotle considers there to be a determinate magnitude for every reality, to the extent that this concept corresponds to the idea we have of the ancient cosmos. However, to find the idea of "right size" in a contemporary biologist's work could be surprising.[28] It raises a question: Does taking into account the scale effect necessarily lead to an order of absolute magnitude? To attempt to answer this question, we should first grasp the theoretical formulation of the scale effect as it is exposed by Galileo in 1638, in the *Discourses and Mathematical Demonstrations Related to Two New Sciences*.

Devoted to the resistance of materials and local movement, the book was written in Italian and uses a process the author employed previously in the 1632 *Dialogue Concerning the Two Chief World Systems*. It takes the form of a dialogue between Salviati, Galileo's spokesperson; Sagredo, who enacts the innocent point of view; and Simplicio, whose role it is to defend Aristotelian tradition—an ungrateful position in this context. The question of size is addressed on the first day. Salviati refers to the technicians' opinion that what is true for a small machine does not necessarily apply to a larger one, noting that increasing the size of an object goes hand in hand with a decrease in solidity. He then extends this observation to trees and animals: "Who does not know that a horse falling from a height of three or four cubits will break his bones, while a dog falling from the same height or a cat from a height of eight or ten cubits will suffer no injury? Equally harmless would be the fall of a grasshopper from a tower or the fall of an ant from the distance of the moon."[29]

Nobody seems particularly bothered by the ant falling from the sky, and Salviati continues his argument: "And just as smaller animals are proportionately stronger and more robust than the larger, so also smaller plants are able to stand up better. . . . Nature cannot produce a horse as large as twenty ordinary horses or a giant ten times taller than an ordinary man unless by miracle or by greatly altering the proportions of his limbs and especially of his bones. "[30]

The dialogue returns to this theme on the second day, when Galileo presents a geometric demonstration of the resistance of "similar cylinders and prisms," and, applying the reasoning to living beings, Simplicio highlights the gigantism of whales. Salviati responds to the objection, referring to the density of water. Though using other words, he is referring to Archimedes' principle of displacement, which made this gigantism possible. Salviati concludes that it is impossible for animals to grow immensely in height "only by employing a material which is harder and stronger than usual, or by enlarging the size of the bones, thus changing their shape until the form and appearance of the animals suggest a monstrosity."[31] That is to say, changing size means changing shape. Galileo illustrates this with a drawing showing "a bone whose natural length has been increased only three times and whose thickness has been multiplied until, for a correspondingly large animal, it would perform the same function which the small bone performs for a small animal."[32] Indeed, three and a half centuries later the physiologist Knut Schmidt-Nielsen notes, in a book on scale effects in animal physiology, that the increase in thickness visible in these illustrations is exaggerated: the big bone is 9 times wider than the small one, whereas a ratio of 5.2 would have been sufficient.[33] From the perspective of the American physiologist, this miscalculation obviously did not detract from Galileo's decisive role in the field.

D'Arcy Thompson highlights the role played by Galileo. This Scottish zoologist, as familiar with Aristotle's work as he is with trigonometry, authors an astonishing book called *On Growth and Form*, filled with ideas as luminous as its editorial destiny was entangled. The first version of the book is published in 1917, in a volume with 793 pages. Thompson publishes a second revised and expanded edition comprising 1,116 pages himself in 1942, after which it is reprinted many times. In 1961 John Tyler Bonner produces a posthumous, abridged, and updated edition about 350 pages long. This edition is published in paperback with a preface by Stephen Jay Gould.[34] These multiple editions spread the author's famous diagrams showing progressive transitions from one form to another.[35]

D'Arcy Thompson devotes the first chapter of his work to scale effects. He recalls Galileo's reasoning and demonstrates its relevance.[36] He praises the clear reasoning of the Genevan physician Georges-Louis Lesage, who is published in 1805 by Pierre Prévost.[37] In this retrospective, D'Arcy Thompson focused on both physiology and morphology. He refers to Jean-François Rameaux and Frédéric Sarrus, respectively a doctor and a mathematician, whose joint work on metabolism, known from a report published in 1838–1839, establishes that since heat loss by radiation is proportional to the surface of an animal, it varies as the square of the linear dimension, while the production of heat by the organism, being linked to the mass, varies as the cube of this same linear dimension.[38] Using this same observation, the physiologist Carl Bergmann concludes, in 1847, that small animals use more energy than large one, which means that they were less adapted to life in arctic regions. Hence the so-called Bergmann's law. John Tyler Bonner, publisher of *On Growth and Form*, states that this law applies only within a species, and that, even if limited in this way, remains controversial.[39] But D'Arcy Thompson is not particularly interested in verifying the principle of ecophysiology. The purpose of the chapter, as summarized by Stephen Jay Gould in his preface, is to make it clear that small and large organisms lived in worlds where different forces predominate because of the relationships between lengths, areas, and volumes.[40] As a result, spatial scale begins to take form, beyond which or below which one cannot even conceive of living beings.

The French philosopher and mathematician Antoine Augustin Cournot made a similar observation a few decades earlier in his work *Matérialisme, vitalisme, rationalisme*. Contrary to what the title suggests, Cournot does not tackle materialist, vitalist, or rationalist arguments on metaphysics, but rather develops his own philosophical analyses of the sciences of matter, life, and reason. On the question of scale, Cournot espouses the view that "pure geometry . . . the dimensions of bodies . . . are only relative." On this same level of abstraction, "there is no such thing as absolute bigness or smallness." Following from this, he explains homotheties as follows: "the same figures can be constructed on an infinite number of scales, in which case it is said that they are similar." The transposition of "these abstract considerations in the order of physical realities" with what Cournot calls "philosophers and scholars" gives rise to "pleasant tales" and "eloquent tirades." He alludes to Pascal and Swift without naming them, first evoking the two kinds of infinities and then recalling that "after having put a Lilliputian in his pocket, Gulliver was himself put in the pocket of a giant." Then, dismissing these fictional stories, he adds:

> But cosmology leads to more positive facts. In our reality, every type of phenomenon occurs on its own scale, and usually we can observe a very abrupt shift from one scale to another. We do not see crystals the size of planets or mountains, and even if we increased the power of our microscopes, we would not find anything like a solar system in a crystal or a drop of water. Likewise, we would not find an oak tree or an elephant in microscopic plants or animals.[41]

The claim that a solar system cannot be found in a drop of water was debunked a few decades later when the representation of the atom emerged with electrons revolving around a central nucleus, like so many small planets around a big star. Today, we know this is a convenient but inadequate representation of atomic theory.[42]

Thus Cournot, like D'Arcy Thompson a few decades later, emphasizes the fact that the various realities have an order of magnitude beyond which we can only imagine their existence in a fictitious way. The French philosopher could have written this sentence by the British naturalist: "Men and trees, birds and fishes, stars and star-systems, have their appropriate dimensions, and their more or less narrow range of absolute magnitudes."[43]

Seeing Galileo's name appear in the history of this discovery, as it appears in the history of the earth's movement, is an invitation to compare the two. The scientific revolution of the seventeenth century seems to have brought about a double rupture. The first, many times celebrated, results in a decentering of the world, while the second, almost unnoticed, assigns an order of absolute magnitude to each domain of reality. The latter is counterintuitive, going against the grain of how from childhood we learn to think about the relativity of spatial magnitude. Although technicians and engineers, trained through their handling of scale models, are familiar with an order of magnitude specific to each reality, geometricians and philosophers sometimes have difficulty accepting it. The French mathematician Henri Poincaré himself offers an illustration in *Science and Method*, which was published in 1908. Devoting several pages to the notion of an absolute space that could exist in itself independently of spatial realities, he asks his reader to imagine that all the dimensions of the universe become a thousand times larger overnight. Our body and all things, including the graduated rulers themselves, will have undergone the same transformation. He deduces that we would not notice anything. This imprudent conclusion is criticized with the objection that pork butchers would see the sausages hanging in their shops fall down.[44] The volume and therefore the weight of these sausages would have been multiplied by a billion, while the strength of the strings, proportional to their section, would have been multiplied by a million. Could Poincaré have been unaware of what Galileo already knew? It is probable that he neglected it, because this supposition was meaningless to him. "In reality," Poincaré concludes, "it would be better to say that as space is relative, nothing at all has happened, and that it is for that reason that we have noticed nothing."[45]

This is a key issue, as it touches the very reality of space. The development of quantum mechanics adds more complexity. In a 1947 in the *Journal de Psychologie*, the French philosopher Pierre-Maxime Schuhl discusses how "contemporary physics" is breaking with the illusion that the laws of physics remain the same at any scale.[46] Schuhl is referring to the scale of microphysics, but insects do not live in a quantum universe. The irreversibility of time, the distinction between object and subject, notions of cause and substance—suitable for classical mechanics and found in everyday life—remain operative.

This pinpoints just how precious the world of insects is, being a reality that could be described as tiny but that remains macroscopic. It offers the possibility of forging fictions that, in addition to their poetic power, provide a multitude of thought experiments, which always lead back to questioning what our world would be like if we were one hundred to one thousand times smaller than we are.

For a moment, let us forget the odd charm of Robert Desnos's gigantic ant. It is more improbable by its size than by its linguistic skills—to transform from a few millimeters in size to eighteen meters would require such a change of structure that it would no longer look like an ant. The giant spiders that the artist Louise Bourgeois delivers to our imagination necessarily have reinforced legs that are made of a completely different material than those of their models.[47] The problem does not only arise in terms of structure and materials. Physiology intervenes as much as anatomy does. The simple maintenance of an internal temperature varies with size. Although Lewis Carroll does not mention it, the 25-centimeter-high Alice would have been shivering, since her body mass would have reduced proportionately. Breathing is also impacted insofar as it involves the ratio between the exchange surface and the mass of the body. In this respect, the existence of giant dragonflies in the Carboniferous period poses a problem for paleontologists. Their gigantism—however relative it may be—seems incompatible with the insect's respiratory system, based on tracheae. They become plausible again when geophysical data and theoretical models suggest that the atmosphere at that time had a higher oxygen content than the current one.[48]

The universe contained in a mite, through which Pascal attempted to give us the sense of the infinite in order to prepare us for conversions, belongs to apologetic rhetoric. And although Swift is, to a certain extent, sensitive to this question of scale, the dwarfs and giants that Gulliver meets are not only strange, they are totally incompatible with the laws of physics.

The illusion that insects live in another world, a world where the laws are different, arises from the fact that although the same laws operate there, they do so on a smaller scale, making their effects different. Unity of the laws produces the otherness of the phenomena and explains the impression of encountering something fantastic.

2

AN INORDINATE FONDNESS
FOR BEETLES

Haldane, the aforementioned British biologist, is said to have met with some theologians one day who asked him what could be concluded about the nature of the Creator from a study of his creation. The story goes that Haldane replied: "an inordinate fondness for beetles." George Evelyn Hutchinson, who reported this event without guaranteeing its authenticity, gave no further details about the circumstances of this interview.[1] Perhaps it was apocryphal. The comment nonetheless translated the biologist's enthusiasm in the face of the huge number of Coleoptera species—300,000 to 450,000 to count today, representing around 40 percent of all insect species. They are easily recognizable by their hardened forewings, called elytra, which cover and protect the hindwings. If we add that they undergo complete metamorphosis between the larval stage and the adult stage, it is not surprising that Coleoptera have long been considered a homogeneous group. However, the internal structure found within this group, like that of other insect groups, is not so straightforward, and its conceptual architecture results from a series of methodological choices.

THE CLASSIFICATION PRINCIPLE

While the name *beetle* can be applied to all Coleoptera, in every-day language it is often reserved for members of the Scarabaei-dae family. In the first instance, we would say that a ladybug is a beetle, since the Coccinellidae family is included in the Coleoptera order. Upon a second look, Scarabaeidae and Coccinellidae are considered to be two of the many families included in the Coleoptera order. In order to fully understand this approach, we have to admit that assigning a designation to a living being amounts to placing it in a conceptual edifice made up of groups nested within each other. Insects are no exception to this rule. Even a nonspecialist, who lacks images and words, may be reluctant to regard ladybugs as a family—that is, as a group occupying the same rank in the *system of classification* as grasses, palms, canines, or felines. An entomologist, on the other hand, like all naturalists, would encompass these successive relations of inclusion in an overall hierarchy, which proceeds from kingdom (animal or plant) to individual via division, class, order, family, genus, species, race, and variety. This particular hierarchy is, in fact, obsolete. The kingdom distinction has been redefined. The term *phylum* replaced *division,* providing a convenient mnemonic: "King Philip Came Over from Great Spain." The notion of race has been criticized, not only for its political uses, but also simply for its lack of scientific consistency. Furthermore, classifications have intermediate categories: subclasses, superorders, subfamilies. Sometimes, tribe is placed between the subfamily and the genus. This word is confusing since it designates a subset of the family in biology, whereas it refers to the opposite in anthropology. One could be tempted to compare, joyfully or sadly, the abundant complexity of current classifications to the reassuring regularity of previous classifications. It would, however, be hard to find a golden age of classification. First of all, from the middle of the eighteenth century to the middle of the nineteenth century, the notions of family and order often designated the same level of classification. Further, the multiplication of intermediate categories is not a recent phenomenon, since Jean-Baptiste Lamarck already complained about it in 1809:

> Some modern naturalists have introduced the custom of dividing a class into several sub-classes, while others again have carried out the idea even with genera; so that they make up not only sub-classes but sub-genera as well. We shall soon reach not only sub-classes but sub-orders, sub-families, sub-genera and sub-species. Now this is a thoughtless misuse of artifice, for it destroys the hierarchy and simplicity of the divisions, which had been set up by Linnaeus and generally adopted.[2]

Indeed, what the author of *Zoological Philosophy* criticized was precisely the flexibility of this conceptual structure. Inserting these extra steps may have destroyed the simplicity of Linnaean divisions, but in doing so it certainly ensured the sustainability of the overall structure. This irreversible hierarchy is not to be confused with a classification in the sense of an ordering of elements that are placed one after the other, from first to last. It is based on inclusion relationships. We could say it is interlocking.

THE PARADOXICAL CROCODILE

In the case of insects, we see the definition of homogeneous groups. Thus in Aristotle we find insects that have two wings, like flies, and those that have four, such as bees, or those whose wings are protected by a sheath as is the case with beetles. While the ancient Greeks relate butterflies to Psyche, the goddess of the soul—Aristotle says they "are produced from caterpillars."[3] Thus groups arose in which today's entomologists retrospectively find Diptera, Hymenoptera, Coleoptera, Lepidoptera, etc. These groups have shown great robustness, since they still constitute the many subgroups included in the insect group. However, this feeling of familiarity fails as soon as we approach the question of limits, in other words, as soon as we try to determine how far the concept of the insect stretches. The term *insect*, along with the Latin term *insectum* and its Greek equivalent *entomon*, which is found in "entomology," refers to "cut into," which evokes constrictions in the shape of an insect's body. The etymology is based on the segmented aspect of these animals. But this aspect, be it common to all species of insects in the current sense, is equally noticeable in spiders, scorpions, millipedes, and even in some worms. This explains the common meaning of the word *insect*, which the general public uses, even today, to designate what naturalists have called terrestrial arthropods since the end of the nineteenth century. At first glance, the popular notion of insect is very close to that used by Aristotle. The Greek philosopher, however, included insects in the more general category of animals he believed to be bloodless, while he gave the segments of which their bodies are composed the property of continuing to live after being cut into.[4]

The etymological definition of insect as a segmented animal appears very broad. However, to the French entomologist René Antoine Ferchault de Réaumur, it seems too narrow when he published the first volume of his *Mémoires pour servir à l'histoire des insectes* (Memoirs to serve as a history of insects) in 1734. He specifies that he would not limit himself to the study of animals "that have incisions" or of animals "that have a certain smallness." To illustrate his point, he does not hesitate to write: "A crocodile would be a curious insect; yet I would have no difficulty in giving it that name."[5] It should be noted that in the following volume Réaumur does not refer to crocodiles or other reptiles. It is nonetheless significant that he considers the possibility of extending the object of his study in this way. He explains it by recognizing that he would "willingly" include in the class of insects "all animals whose forms do not allow us to place in the class of ordinary quadrupeds, in that of birds or in that of fish."[6] We see that contrary to contemporary classification requirements, Réaumur accepts taking into account a group obtained by subtracting one or more subgroups. Moreover, Réaumur does not hide the fact that these classification problems are not the ones that interested him the most: "I have already stated enough that the part of the history of insects in which I am most interested is the one that looks at their genius, their industry."[7] He likewise writes the following in a memoir that remained unpublished for a long time: "The different species of ant exhibit few remarkable differences in form. When, therefore, the external appearance of one ant is known, that of all other species is also fairly well understood. . . . It is by means of their habits and their different proclivities that it is often easier and always more agreeable to distinguish them from one another."[8]

Obviously, Réaumur finds naming and classifying the multitude of insects from their morphological description to be neither appealing nor essential. He prefers to focus on the study of behavior. On this point, his younger rival, the scholar Georges-Louis Leclerc, Comte de Buffon, agrees in 1749, when he wrote about classifying quadrupeds according to their limb anatomy:

Is it not better to arrange, not only in a treatise on natural history, but even in a picture or elsewhere, objects in the order and in the position in which they ordinarily appear rather than to force [them] to be together under a supposition? Is it not better to follow the horse, which is soliped, by the dog, which is fissiped, and which customarily follows the former, than by a zebra which is little known to us, and which perhaps has no other relation with the horse than to be soliped?[9]

On the other hand, this disdain for classification—we could almost use the word *disgust,* considering the level of subjectivity—opposes the two French naturalists and their Swedish colleague, Carl Linnaeus. For the latter, describing, naming, and classifying minerals, plants, and animals was one of the major objectives of natural history. Moreover, it is important to distinguish two aspects of Linnaeus's work that are complementary but had a different fate: nomenclature and classification.[10]

Linnaeus introduces the so-called Linnaean nomenclature against his own will. For a long time he remains attached to lengthy descriptive names. But the long names are inconvenient, and it is simpler to designate each species by means of a binomial formed by the genus (which may be common to several species) and a specific epithet. Thus, in botany, white clover becomes *Trifolium repens,* and red clover *Trifolium pratense.* Similarly, in zoology, the great tit is named *Parus major* and the titmouse *Parus palustris.* Botanists thus designate plant species by the name attributed to them by Linnaeus himself in *Species plantarum* (1753) or given to them by another botanist since that date, provided that it follows Linnean rules. Zoologists do the same, but their starting point is the tenth edition of *Systema naturae* (1758). This nomenclature is all the more useful for entomologists as the species they study are so numerous.

Botanical classification is based on a system of twenty-four classes, with classification depending on the number and appearance of the sexual organs. Zoological classification, as Linnaeus presents it, comprises six major divisions: quadrupeds, birds, amphibians (which then included reptiles), fish, insects, and worms. Use of the term *worm* during Linnaeus's time was quite different than its use today, since it applied not only to worms but also to animals with shells and to starfish. From one edition to another, Linnaeus introduces changes. In the fourth edition, he places whales and dolphins among amphibians; in the tenth, he adds them to viviparous quadrupeds in the class of mammals. Not surprisingly, he uses the term *insect* for all animals that could be considered retrospectively as arthropods. They include seven orders whose names are now familiar to any entomologist, seasoned and beginner alike. Indeed, we find the aforementioned Coleoptera, along with Hemiptera (to which bedbugs belong), Lepidoptera, known to the world under the name of butterflies. Diptera (flies and mosquitoes) and Hymenoptera (wasps, ants, bees, etc.) are distinguished by the possession of a pair or two pairs of wings. Neuroptera includes species that still belong to this group, such as the Fourmilion, but also others that appear today in the order of Odonata (dragonflies and damselflies). Finally, under the term *Apteres* (that is to say, without wings), Linnaeus arranges insects that have always been considered to be insects, such as fleas and lice, but also spiders, ticks, and even all our current Crustacea.[11]

We could say that from the middle of the eighteenth century the history of the concept of insect is a series of exclusions that gradually tighten its scope, until it only applies to animals with an external skeleton, having a head, a thorax, and an abdomen, three pairs of legs, two pairs of wings (possibly atrophied or missing), and a pair of antennae.[12] This specifically excludes spiders, centipedes, and woodlice.

DRAWING LINES

There are lively debates on these topics in Paris in the middle of the French Revolution, echoed in the minutes of the meetings of the Natural History Society of Paris, recently published by a historian specializing in that period.[13] Thus, on 21 Floreal year 3 according to the revolutionary calendar (May 10, 1795), the entomologist Pierre-André Latreille reads "a memoir in which he makes it felt that it is useful and easy to make a distribution of insects according to the consideration of the mouth."[14] The Danish entomologist Johann Christian Fabricius has suggested such a classification. Latreille agrees that resemblances in mouthparts usually lead to resemblances in other parts of the body, but by qualifying this correlation he renounces using a single criterion as the basis for classification and adheres to the idea of classification by natural families. A discussion then begins "as to the manner in which woodlice should be categorized" and on crustaceans in general.[15]

This discussion reminds us just how thorny determining what can be considered an insect is. Where are the lines to be drawn?

In 1800, year 8 of the French Republic, Lamarck has held the Natural History Museum chair of zoology for "worms and insects" since 1793, which he first designates as "animals without vertebrae" or "invertebrates." In the opening speech of his zoology course, published the following year, he expresses himself in these terms: "The Class of Crustaceans, that is to say the second class of animals without vertebrae, the one that includes animals that had until now been confused with insects, because they have articulated legs and antennae like insects; this class, I say, must immediately follow that of molluscs, and it is no longer permissible to confuse the animals in it with those who truly deserve the name of insect."[16]

In the end, this very diverse set, which includes crabs, lobsters, spiny lobsters, shrimp, and crayfish, along with daphnia and woodlice, is admitted as a class. However, cirripedia (barnacles) remained separate; they would not be included in the crustacean class until the nineteenth century. However, the most striking distinction is the one Lamarck establishes between insects and arachnids. The latter form a class that includes, as it does today, spiders,[17] harvestmen, mites, and scorpions; however, unlike today's arachnids, Lamarck's arachnids also include millipedes. Lamarck characterizes insects by their anatomical characters,[18] such as the presence of three pairs of legs, and by physiological characters such as going through metamorphosis during development, whether it be complete or incomplete.

All these reorganizations could give an impression of arbitrariness, and Lamarck himself considers classes, orders, families, genera, and species as "means of our invention" that we cannot do without "but that must be used with discretion" and by "subjecting them to agreed-upon principles."[19] For that matter, Lamarck recognizes that in order for zoology to classify animals it must borrow from comparative anatomy. On this point, he implicitly agrees with Georges Cuvier. This agreement underlines the importance of the topic, just as it confirms the idea defended by Charles C. Gillispie that from 1800 to 1830 the naturalists at the Museum of Natural History saw beyond their oppositions and constituted a true "scientific community" with "research programs."[20] The expressions used by the American historian are deliberately anachronistic, but they do reflect the vitality of the Parisian scientific milieu at the beginning of the nineteenth century and the theoretical importance of the problems that were dealt with at that time.[21]

THE NATURAL METHOD

Among the issues at hand, constructing a natural classification occupies a central place. It is found at the heart of *Considérations générales sur l'ordre naturel concernant les classes des Crustacés, des Arachnides et des Insectes* (General considerations on the natural order concerning classes crustaceans, arachnids, and insects) that Pierre-André Latreille publishes in 1810. Significantly, the book was dedicated to Cuvier, although that does not prevent the author from expressing his friendship for Lamarck.[22] Like Lamarck, he gives arachnids the status of a class, in which he also includes the *Millepieds*. Like Lamarck, he criticizes his contemporaries' tendency to divide classes into "subclasses," orders into "suborders," and genera into "subgenera," thus multiplying taxonomic levels. As for the construction of a natural classification, Latreille summarizes it as follows, that the "distinctive notes of objects are taken from all their parts."[23] He expounded on this principle in 1795 at the Natural History Society when, in reporting Fabricius's work, he points out that the mere consideration of the mouthparts sometimes leads him to put "in the same order, and genera, species which had no other relation than this resemblance in the form of the mouth."[24] The same conviction animates Cuvier when, responsible for reporting to Napoleon I on the progress of the natural sciences, he approaches the subject of entomological classification. After discussing Jan Swammerdam's classification based on the characteristics of metamorphosis, Linnaeus's "number and texture of wings," and Fabricius's approach based on "the organs of manducation," Cuvier concludes: "The truth is that it is necessary to combine these three kinds of characters to arrive at something natural."[25] Yet, however close to the natural order, a classification remains an intellectual construct. The classifier's dream, then, is to arrive at a classification that would be natural in the fullest sense, that is, in the sense that its naturalness would prevail over its constructed character. At a minimum, naturalness can be marked by scholarly consensus. To take a very simple example, dandelions have yellow flowers and chicory has blue flowers, but in both cases what is commonly called their flowers are actually inflorescences composed of many small flowers. In contrast, buttercups are simple flowers. It is easy to understand that, in order to classify plants, a botanist prefers to place dandelions closer to chicory than to buttercups, which means that he will naturally give more weight to structure than to color. By thus evaluating the respective weight of the characteristics in half a dozen "families" recognized by all the authors (such as gramineae, compounds, umbellifers, and leguminosae), it becomes possible to form dozens of other families based on the same model so as to categorize the other plants. This is the approach taken by Antoine-Laurent de Jussieu in *Genera plantarum* (1789), and Cuvier admires this implementation of the natural method. At the same time, he considers that resemblances are "more striking, and their causes easier to find" in animals than in plants, so that, according to him, the natural method would rapidly impose itself in zoology as well as in botany.[26] On this point, the development of zoology, and more particularly of entomology, was to prove Cuvier right, but this triumph brings with it a disruption in naturalistic disciplines, which results in questioning the fixity of species to which Cuvier is attached. We can qualify this upheaval, without exaggeration, as a revolution, the Darwinian revolution, that will result in a profound change in the conception of how to classify the living.[27]

DARWIN AND DESCENT
WITH MODIFICATION

We find the essence of this revolution in chapter 13 of *On the Origin of Species*. Charles Robert Darwin finds that while the classification of living beings is "not arbitrary like the grouping of the stars in constellations, the existence of groups would have been of simple signification, if one group had been exclusively fitted to inhabit the land, and another the water."[28] Faced with the question of the meaning of classification, some naturalists "look at it merely as a scheme for arranging together those living objects which are most alike, and for separating those which are most unlike" or a way to shorten descriptions: to describe a dog, for example, it would suffice to add to a proposition giving the features common to all mammals, another giving the traits common to carnivores, and then to add a sentence on what distinguishes dogs from other carnivores. This type of answer does not satisfy Darwin, who notes: "The ingenuity and utility of this system are indisputable. But many naturalists think that something more is meant by the Natural System; they believe that it reveals the plan of the Creator; but unless it be specified whether order in time or space, or what else is meant by the plan of the Creator, it seems to me that nothing is thus added to our knowledge."[29]

To put it in words other than his own, logical analysis is inadequate and a theological response is sterile. As a result, the meaning of classifications must be sought not in what naturalists think but in the rules they give themselves to build the classifications.

As Linnaeus said, it is the genus that gives the characters, and not the characters that make the genus, reflecting the idea that classification contains something more than resemblance.[30] Darwin proposes that this something more, this hidden link, could be found in genealogical proximity.[31]

Moreover, naturalists have been applying this principle for a long time, as they have placed individuals of both sexes, or larvae and adults, in the same species even when they take very different forms.[32] Reviewing the main methodological rules of classification, Darwin goes so far as to say that common ancestry is the link that naturalists have sought unconsciously through the various principles by which they have regulated their work.[33] This is the manner in which he explains how, in classifications, they attach so much importance to rudimentary organs, a vestige without physiological utility: "Rudimentary organs may be compared with the letters in a word, still retained in the spelling, but become useless in the pronunciation, but which serve as a clue in seeking for its derivation."[34]

Entomology provides Darwin with a significant example. On small windswept islands, natural selection tends to reduce Coleoptera wings to rudimentary organs. In this type of environment, Coleoptera are likely to be washed away by the wind, to the point that it is more advantageous for them not to fly. Here Darwin relies on the work of Wollaston in Madeira.[35]

For Darwin, the expression of genealogy is what distinguishes the classification of the living from a simple artificial classification of arbitrary objects. This principle had hitherto remained unknown to the naturalists themselves, but, unbeknownst to them, it underpinned their methodology.

A METHODOLOGICAL REVOLUTION

Above all, the upheaval brought by Darwin concerns the interpretation of systematics while hardly affecting its daily practice. However, a real methodological rupture came with the advent of phylogenetic systematics—also called cladism or cladistics—initiated during the second half of the twentieth century by the German entomologist Willi Hennig and then extended to all animal and plant organisms.[36]

To understand the basics of phylogenetic systematics, it is first important to dispel frequent points of confusion. During the second half of the twentieth century, the practice of classification witnesses two innovations that profoundly renew it. One is cladistics, and the other is taking into account molecular characters. In fact, these two innovations are often linked, but they are conceptually distinct: systematists can take molecular characters into account and still remain skeptical about the validity of cladistics; conversely, cladistic reasoning can be applied to material that does not lend itself to molecular analysis, such as fossils. This is one of the points that Claude Lévi-Strauss underlined in his lecture devoted to Guillaume Lecointre and Hervé Le Guyader's phylogenetic classification of life:

After Hennig's founding work, the position of cladistics in the life sciences was complicated by the collaboration (or perhaps occasional competition) that emerged between morphology and molecular biology. The latter has become increasingly important because, as it is situated at a deeper level of life, one may wonder whether it is not what ultimately holds the keys to true phylogeny. Yet it does not give simple answers either. As with morphology, one has to choose between possible trees.[37]

We note that the terms cladism and cladistics—practically synonyms—come from the Greek *klados*, twig, spinning the metaphor of the family tree. But if cladists are not the only ones using tree diagrams, they are different in the use they make of them and in the rules they formulate as to their construction. Without claiming to detail this use and these rules, one can grasp the guiding principles by referring to an article by Willi Hennig, entitled "Phylogenetic Systematics" published in 1965 in an entomology journal.[38]

The central point is "the division of the concept of resemblance into various categories of resemblances."[39] This division is established when we introduce the notion of convergence, which indicates a resemblance resulting from the adaptation of two organisms to similar environments, but which should not be taken into account when grouping the species into daughter units. For example, to bring whales and fish together, because of their resemblance in external form, would only result in the establishment of a polyphyletic group, that is to say a group that derives from several ancestors. This example is not in Hennig, because no naturalist would waste time discussing a category that would group fish and whales, but it serves the purpose of skirting the limits of resemblance. However, says Hennig, "even when purged of convergence," resemblance does not provide exclusively monophyletic groups—that is, formed from the descendants of a single ancestor—which is required by the idea of phylogeny.[40] By introducing the notion of pleomorphism to designate the initial state of a character, Hennig explains: "This arises from the fact that characters can remain unchanged during a number of speciation processes. Therefore, it follows that the common possession of primitive ('plesiomorph') characters which have remained unchanged cannot be evidence of the close relationship of their possessors."[41]

Having chosen to name the initial state of a character as "ple-siomorph," Hennig calls the state derived from this same char-acter "apomorph." Only a resemblance based on sharing this derived state is the sign of a close relationship. In the opposite case, if we "associate species" because they present "plesio-morph" characters, we obtain a "paraphyletic" group, that is to say a group that includes an ancestral species and only a part of its descendants.[42]

Hennig applies these concepts to the insect class, then divided into two subclasses: Apterygota and Pterygota. Until recently, the first category included insects that had no wings, and whose ancestors did not have any. The second category included but-terflies, beetles, cockroaches, grasshoppers, dragonflies, bees, wasps, flies, mosquitoes, bedbugs, etc. In other words, it included all winged insects, including those that only have wings at the time of a nuptial flight—this is the case of the ants—and even those, such as fleas and lice, whose ancestors one can assume had wings.

The crucial point is that one of the two subclasses, the Apter-ygota, is a paraphyletic group: "The sole common ancestors of all the so-called 'Apterygota,' for instance, were also the ances-tors of the Pterygota, and the beginning of the history of the Apterygota was not the beginning of an individual history of this group, but the beginning of the individual history of the Insecta, which were at first Apterygota in the morphological-typological sense."[43]

The evolution of the discipline has confirmed Hennig's anal-ysis, as evidenced by an article on Collembola, recently published in the Friends of the National Museum of Natural History newsletter. The author, Jean-Marc Thibault, confirms that the old group of Apterygota was based on a shared primitive char-acter, the absence of wing. He adds that the six-legged species, the hexapods, are now divided according to the position of their mouthparts. Insects, strictly speaking, have their mouthparts on the outside of the head. They include the ancient Pterygota, to which the Thysanura have been added.[44]

So we see, the concept of insect, which for Réaumur extended all the way to the crocodile, is confused with arthropods almost through to Linnaeus; Lamarck detaches the arachnids from the concept of arthropods, while Latreille subdivides it into natural families, and today the concept is still undergoing a change in scope.

WHERE WERE THE INSECTS
BEFORE THEY EXISTED?

Logicians have long used the rule that the scope and comprehension of a concept—in short, the realities to which it applies and the characteristics that define it—vary in opposite directions. This means that the broader the definition becomes, the less it concerns things. This rule applies here in the case of classical systematics.

Let's keep in mind that for a long time the class of insects had been divided into two subclasses: Apterygota (nonwinged insects) and Pterygota (winged insects). Note that the definition of Pterygota included one more trait than insects, insofar as Pterygota were defined by the presence of wings. Note also that Pterygota applied to a smaller number of objects. Yet this presupposes that one considers both the absence and the presence of wings to be a crucial character.

If we now take a cladist perspective, where is the inverse correlation between comprehension and extension? To the extent that that the existence of wings, for example, is a positive characteristic, in the logical sense, while their absence is a negative characteristic, and given the cladists' penchant for not using negative characters, it seems difficult to consider the winged/ wingless alternative as a defining element. However, the cladist rejection of negative characters supposes that we can always distinguish between the affirmation of a negative character and the negation of a positive character.

Another difficulty arises when reading ancient entomological texts, one that prompts us to engage in verbal contortions, speaking of "animals designated as insects and that we would consider today simply as terrestrial arthropods." For example, how can we talk of work dealing with spiders prior to 1800? Should we remember that they were then considered insects and present this as an error that could have been corrected by a better understanding of what an insect is? Should we, on the contrary, assume that an insect is simply what each era designates as such? In the first case, do we not risk making current knowledge the absolute norm? Were we to adopt the second solution, would we not slide toward a relativism that would consider all classifications as equal and would suggest that they are all arbitrary? Rather, should we not adopt a conditional frame of reference and express ourselves thus: "if we define an insect in such and such a manner, then this organism is, or is not, an insect?" This leaves open the possibility that some definitions are better than others.

These difficulties come from the fact that insects appear as natural beings perceived through the filter of culture. We do not easily forget that these creatures are of nature due to the behavior of some that escape our control and cause inconvenience or diseases, or the unexpected beauty of others, or their aggressive proliferation, or their indiscreet familiarity. That they are at the same time perceived through the filter of culture manifests in the multiple ways in which they have been defined, identified, described, named, classified, and circumscribed by entomology, and even more so in the multiple images drawn from literature or the arts.

Here the history of science joins the history of art. Albrecht Dürer's 1505 stag beetle is known for its realism.[45] However, examples of insect representations prior to the seventeenth century are relatively rare. Hence the particular interest of a recent study by entomologist Colette Bitsch devoted to the illuminations found in a fourteenth-century Italian manuscript. These illuminations, made by an anonymous artist at the request of rich Italian merchants, the Cocharelli, were made with great precision, suggesting that the artist engaged in real observation, making the identification of the species represented possible. In addition to the aesthetic quality of these images, their accuracy proves to today's entomologists the range of implicit knowledge held by the entomologist of the past.[46]

3

AN ENTOMOLOGIST'S
POINT OF VIEW

I n the courtyard of a private mansion, a bee approaches a potted orchid, while an aging baron and a young shopkeeper indulge in games of coquetry and seduction. Hidden behind a shutter, the narrator observes them, just as he had planned to observe the insect and the plant. This scene is found in the first pages of Marcel Proust's *Sodom and Gomorrah*.[1] A little further on, Proust refers to Darwin's work on flower fertilization, comparing the baron's attitudes to a flower's corollas used to attract pollinating insects.[2] In the same pages, the narrator is astonished that although he used to find jellyfish abhorrent, he is interested by Michelet's description of them.[3] The appearance of natural history in *The Search for Lost Time*, however discreet, does not escape notice by certain commentators. In 1924 the French literary critic Henri Massis attacks André Gide's immorality in the name of traditional values, opposing him to Proust, "who analyzed the worst aberrations with an objectivity similar to that of an entomologist observing insect mores."[4] This comparison to an entomologist seems self-evident and can be found on other occasions. The philosopher Henri Gouhier reports that Paul Moultou, who published Rousseau, defended him against Voltaire, adding: "One imagines the great man throwing on his young opponent the look of an entomologist."[5] Here the allusion to entomology is reduced to the idea of astonished coldness, but it often goes further, implying sharpness and rigor in observation.[6] On September 25, 2009, *La Libre Belgique* asserts that Balzac "looked upon people as an entomologist would" in "At the Sign of the Cat and Racket." This type of expression is common enough in French for a television channel to call Claude Chabrol "the entomologist" in the title of a show paying homage to the filmmaker known for his depictions of human psychology. Occasionally, an adjective is added, transforming the entomologist into "cruel" or "compassionate."

WRITERS AND ENTOMOLOGISTS

Why mention both writers and entomologists together? Entomologist writers are not all that numerous, although they do exist. At the beginning of the nineteenth century, Charles Nodier is a member of the French Entomological Society, and he writes fantasy stories that are still quite enjoyable to read (particularly *La Fée aux miettes*).[7] A century later, Vladimir Nabokov, renowned for *Lolita* (1955), describes a passionate relationship between a brother and sister in *Ada* (1969). He makes frequent allusions to entomology in *Ada*, which is hardly surprising for an entomologist of such stature as to manage the butterfly collection at Harvard University's Museum of Comparative Zoology between 1942 and 1948. However, there is no sense looking for any of the novelist's anticonformity in his practice of entomology. Stephen Jay Gould convincingly demonstrates that Nabokov is competent enough in distinguishing and classifying similar species and yet never proposes any novel interpretations. Only his "passion for accuracy" could be found in both his science and his literature.[8] In a completely different context, the German novelist and essayist Ernst Jünger's entomology, as anecdotal as it is in comparison to the man's nationalism and bellicosity, contributes to his literary celebrity.[9]

Outside of these special cases, speaking about entomology when discussing an author is akin to suggesting that he or she observes humans as if they were not human. The comparison is understandable, yet one has to question why insects are always the ones used in the comparison. Why not compare, say, to an ornithologist or to a microbiologist? Perhaps one element of a response can be found in Jean-Henri Fabre's *Souvenirs entomologiques*, published between 1879 and 1907. Fabre spends several pages describing himself observing insects and depicting himself as a country bumpkin, a young school teacher, an unlucky inventor, and, finally, an independent scholar supporting his family with his book royalties.[10] Anecdotes and autobiographical comments alternate with entomological descriptions and observations. Some pages read like discovery narratives,[11] such as those consecrated to great peacock moths (aka giant emperor moths), when a female sheds its cocoon in his laboratory: "Well, on the morning of the sixth of May, a female Great Peacock Moth came out of her cocoon in my presence, on the table of my insect-laboratory. I at once caged her under a wire-gauze bell-jar. I did not think much about the matter. I kept her on general principles, for I am always on the lookout for something to happen."[12]

Around nine in the evening, Fabre continues, one of his children is rushing about crying out that the house was invaded by moths as "big as birds." Fabre recalls the prisoner he held under a bell jar. He wonders how these "forty lovers who have come to pay their respects" had found their "princess." Was it by light, sound, or odor? It was not by light, as "darkness was light enough for him." Nor could it be sound, as Fabre notes that the creature was silent. Yet, he notices that the influx of males stops when he encloses the female in a hermetically sealed container. He suspects that "emanations analogous to what we would call odors, extremely subtle emanations, completely imperceptible to us, and yet capable of impressing a more skilled olfactory sense than ours."[13] The American historian Frank Egerton notes that the English naturalist had already suspected the action of these imperceptible smells,[14] which correspond to what the biologist A. Bethe studied in the 1930s under the name of "ectohormones" and what biochemist Peter Karlson and entomologist Martin Lüscher named "pheromones" in 1959.[15] The success of this notion of pheromones, which has spread beyond entomology, gives an interesting although retrospective view of this flirting scene that got the whole house excited. Here scientific observation joins literary endeavor.

Although mostly forgotten today, at least in France, Fabre did have a considerable reputation at the beginning of the twentieth century. Once again Proust provides a testimonial in his letters to Jean Cocteau. When Proust expressed his distress over a woman of the world not understanding his work, Cocteau replied, "You are asking insects to read Fabre."[16] The comparison is not limited to aristocratic circles, as demonstrated by the character of Françoise, who is the chambermaid in *In Search of Lost Time*. She expresses limitless devotion to her descendants, while being incredibly harsh with the kitchen wench, to such an extent that Proust makes the following comparison:

> There is a species of hymenoptera, observed by Fabre, the burrowing wasp, which in order to provide a supply of fresh meat for her offspring after her own decease, calls in the science of anatomy to amplify the resources of her instinctive cruelty, and,

having made a collection of weevils and spiders, proceeds with marvellous knowledge and skill to pierce the nerve-centre on which their power of locomotion (but none of their other vital functions) depends, so that the paralysed insect, beside which her egg is laid, will furnish the larva, when it is hatched, with a tamed and inoffensive quarry, incapable either of flight or of resistance, but perfectly fresh for the larder.[17]

Clearly, Proust had read Fabre and knew that the various wasp genera belonged to the order of Hymenoptera. One could think that an entomologist with a rural background like Fabre would write a text more sober and with less emotion than that of the Parisian writer. However, reading *Souvenirs entomologiques* proves the contrary. When Fabre is still a budding entomologist, he reads Léon Dufour, the master who introduces him to the parenting behavior of the *Cerceris bupresticide*.[18] It is clear that Fabre had no fear of literary flourish with chapter titles such as "A Knowledgeable Killer" (chapter 5) and "Three Stabs" (chapter 7),[19] descriptions such as "cruel as an ogress . . . she feeds only on living creatures" (chapter 5), and such phrases as "hungry murderers who lie in wait" (chapter 7). This rhetoric could be interpreted as an artifice linked to him writing for a lay public. But then it is surprising to find it present, although more discreet, in the article Fabre published in 1855 in the *Annales des sciences naturelles*, where he was communicating with scientists on his "Observations of Cerceris and the Causes of Long-Term Preservation of Coleoptera Supplied to Their Larva." Fabre observes the insect digging a tunnel, choosing the right sheltered spot set up like a natural peristyle. It is a "marvel to see how these hard-working miners maneuvered."[20] "Brawls" are frequent. Fabre gives the predator prey and watches as the insect "surveyed around its home for a few moments," seeing the gift, but then flying off without even "honoring" it with a "strike of the jaw."[21] On the following page, he sees the "murderer's abdomen slide under the weevil's belly, curl, and swiftly dart its venomous stinger two or three times" into its victim. He provides an impressive description of the "assassin's formidable talent."[22] This is an example of scientific observation being cloaked in the language of an adventure novel, something found often in *Souvenirs entomologiques*, in Fabre's descriptions of insects *stricto sensu*, as well as of other arthropods such as spiders and scorpions.

SWASHBUCKLING ADVENTURE

It is a spider—a black-bellied tarantula—that Fabre imagines fighting bees and "other wearers of poisoned daggers." The battles are equal, but, in the end, always fatal for one of the opponents. "For the poisonous fangs of the Spider the Wasp has her poisoned dagger or sting. Which of the two bandits shall have the best of it? The Tarantula has no second means of defense, no cord to bind her victim, as the Garden Spiders have."[23]

Fabre does not hesitate in his experiments to provoke battles among arthropods. Thus the black-bellied tarantula faces off with bees, which he chooses from among the largest of their species (*Bombus hortorum* and *Bombus terrestris*).[24]

Often Fabre limits his intervention to placing the opponents in the same place, but he also goes further on occasion. To observe a predatory Hymenoptera of the genus tachyte fighting with a praying mantis, he deprives "the huntress of her prey" and immediately gives it "in exchange, a living mantis of about the same size." That way he can, in his own words, "witness the tragedy."[25]

The praying mantis, a predator's victim here, is a formidable predator as well. Fabre's description of its copulation would not be out of place in a novel by the Marquis de Sade. Biologist Pierre Douzou describes it as "spine-tingling" cruelty. The insect's mating ritual also appears in French symbolist poet and novelist Remy de Gourmont's *Natural Philosophy of Love* and sociologist, writer, and literary critic Roger Caillois's essay specifically on the praying mantis that was published in *Minotaure* in 1934.[26]

Fabre's description records as objectively as possible the positions and movements, but also adds completely subjective interpretations. There is nothing difficult in noting that the male jumps onto the back of his "powerful companion," or observing the length of the "foreplay," but how exactly can the author judge the male's "passionate gaze?" In any case, the passion is fatal, and Fabre affirms that on the day after copulation, "the following day at the latest," the female eats the male.[27] Curious to "know how a second male would be received by the female that had just been fertilized," Fabre observes "the same praying mantis use up to seven males" in a two-week period of time, which he summarizes with this description: "To all, she opens her legs, and all must pay for their nuptial bliss with their lives."[28]

He then describes cases when the female devoured the male's head during copulation—a behavior that has since been given neurobiological explanations—and feigns indulgence in order to better express his emotion: "Eating one's love after consummating the marriage, making a meal of the exhausted dwarf, no longer good for anything, is understandable, to a certain extent, in insects with little sentiment; but to bite into it during the act, that exceeds anything that an atrocious imagination would dare dream."[29]

More recent research shows that some males escape this fate and that praying mantises practice cannibalism outside of copulation.[30]

No less cruel, but more intimate, is scorpion courtship, which leads to the most astonishing narration.[31] It begins with playful games, partners looking, couples forming. "It is all superb in its tenderness and simplicity. The dove is said to have invented the kiss. But I know that he had a forerunner in the scorpion."[32] Yet all the flirting comes to a tragic end. Fabre suspects the "matron" is eating the male, and, to find out, he carefully marks the shelter the couple has found. He recounts, horrified, that after marking "the tiles under which the couples take refuge in the evening after their stroll. What do we find the next morning?. . . I often indeed find the female nibbling and relishing her deceased mate."[33]

Could this be an original discovery of a genetically explicable phenomenon? Or a gross generalization based on singular behavior? Or perhaps even an observation biased by captivity?[34] Whatever the case, these descriptions are among the pages most frequently cited from *Souvenirs entomologiques*.[35]

COMEDY OF MANNERS

Not all of Fabre's insects are belligerent swordsmen or man-eaters, and many scenes could have been pulled from a realistic novel or a comedy of manners.

This is the case of the chapters on the Halictidae. Without forming colonies like honeybees (*Apis mellifera*), one of the Halictus species group together their nests and live with their mother in the burrow where they were born:[36] "Underground, there were ten or so cells . . . the household consisting of ten or so sisters. . . . So, which of the survivors, all having equal rights, will inherit the home?"[37]

All the conditions exist for the bees to battle over the maternal inheritance, but they all use the mother's house without dispute: "They come and go peacefully through the same door, attend to their business, pass and let the others pass. Down at the bottom of the pit, each Bee has her little home, a group of cells which she has dug for herself. Here she works alone; but the passage way is free to all the sisters."

As for the bee who gave birth to the others, she "mounts guard" at the entranceway. Fabre highlights: "She is in fact the foundress of the establishment, the mother of the actual workers, the grandmother of the present grubs. When she was young, three months ago, she wore herself out making her nest all by herself. Now she is taking a well-earned rest, but hardly a rest, for she is helping the household to the best of her power."[38]

Heirs who agree to live together, a grandmother participating in family life—this life scene exudes peaceful happiness. A similarly hard-working but more hectic life can be found among the dung beetles. Currently, this name is used to designate Coleoptera that find their food in animal excrement, especially in cow dung. One species of dung beetle is the sacred scarab worshiped by ancient Egyptians. Fabre describes them transporting their supplies in spherical "pill" shapes. Producing and then carrying the balls is not an easy task. Not only does the beetle have to push it little by little, being careful not to lose it, but it also constantly has to beware of helpers "under the pretense of giving a helping hand, harbor the plan to steal the ball at the first opportunity." Sometimes the beetles have their food stolen by clever neighbors. Sometimes there is even a further heist, by a third thief who robs the robber. This inspires Fabre: "I ask myself in vain what Proudhon introduced into Scarabaean morality the daring paradox that 'property means plunder,' or what diplomatist taught the Dung-beetle the savage maxim that 'might is right.'"[39]

FABLES

Fabre does not forget his poetry. He dedicates a chapter to La Fontaine's fable "The Cicada and the Ant."[40] In the story, borrowed from Aesop, the cicada has done nothing but "sing all through the summer" and with the arrival of winter has no food stores. It asks the ant—the ants in Aesop—to lend it some food. The ant refuses and scolds the cicada for its lack of planning. To a moralist, the fable raises the question of whether artistic activity should be paid for; to an entomologist, it improbability lies in the protagonists' life cycles. Réaumur raises this point in his *Natural History of the Ants*: "Still the charming fable of the ant and the cicada is none the less instructive, though it is certain that the ant knows nothing about storing provisions during the summer and though every year all the cicadas die long before the advent of winter."[41]

The paradox is that in criticizing La Fontaine Réaumur lends himself to retrospective criticism: although it is certain that cicadas die before winter, it is also certain that some ants collect supplies. Réaumur does not know about harvester ants, such as *Messor barbarus*.[42] Other naturalists of his time share his ignorance. The observations of ancient authors that inspired Aesop and later La Fontaine were made in the Mediterranean Basin, where harvester ants live. In contrast, the skepticism of modern readers is that of North European naturalists who have not been in contact with Mediterranean fauna. This geographic determination of behavior, itself a source of divergent opinions, is highlighted in the nineteenth century by an American entomologist named Johann Traherne Moggridge.[43]

Unlike other insect memoirs, the *Natural History of the Ants* was not published during Réaumur's lifetime. Discovered in the mid-1920s by the American entomologist William Morton

Wheeler, it had not yet been published when Fabre devoted a chapter of his *Souvenirs entomologiques* to La Fontaine's fable.[44] According to him, the story was "slander." As a moralist, he stigmatizes the ant's egoism and defends the cicada. As a naturalist, he is anxious to restore the scientific truth: in the heat of summer, when the cicadas play well-diggers and pierce the bark of the trees, it is the ants who come to drink of the sap that flows; and finally, after a few weeks, when the cicadas die, the ants feed on their remains. His analysis includes a poem a hundred verses long in the Provençal language, entitled "La Cigalo e la Fournigo," which Fabre fictitiously attributes to one of his friends. He uses the same artifice in connection with a fable by Florian, devoted to crickets,[45] whose moral is "If you want to live happy, live hidden from sight."[46] Again, Fabre does not just criticize the fable but proposes a new version, supposedly more true.

INSECT PROFESSIONS

These detailed critiques highlight the singularity of the fable, a well-defined literary genre that is both popular and scholarly. In contrast, the innumerable anthropomorphic metaphors that dot *Souvenirs entomologiques* are fragmentary narratives that most often attribute to insects trades with clear social significance.[47] We have seen that under Fabre's pen a fight between a wasp and a spider becomes a duel that has nothing to envy compared to those found in swashbuckling adventure novels.[48] In addition, the author of *Souvenirs entomologiques* constantly assigns "trades" to insects as part of a systematic use of metaphors that was well analyzed by Donald H. Lamore, the author of a thesis on stylistics, published in 1969, called *L'image chez Fabre* (Image in Fabre's work).[49]

Predators appear as bandits and thugs; other insects perform honorable jobs. We see a gatekeeper at the entrance to a bee burrow of the halictidae family. From Fabre's perspective, insects form a world full of craftsmen who are courageous, hardworking, and rather individualistic. Pushed to an extreme, this social approach ends up opposing the mass of laboring insects to an idle minority. For Fabre, ground beetles "cloaked in flashy metal" are only good for making a "feast" of a snail. Cetonia, which seems to come directly from a jeweler's box, spends its time dozing in the heart of a rose. All these big insects are beautiful, but "these beauties do not know how to do anything, they have no industry, they have no profession," and Fabre, in a burst of anthropomorphic populism exclaims, "Long live the modest! Long live the little ones!"[50] Fabre who was considered by one of his biographers as the "Homer of Insects," does not shy away from imitating the eloquence of an orator.[51]

The metaphorical use of the notion of profession is not confined to the sociopolitical register, but also applies to the economic functions these insects fulfill: the dung beetle, which makes droppings disappear, and the burying beetle, which handles corpses, certainly attract less attention than the "blood-drinking cousin" or "the wasp, irascible swordsman with a poisoned dagger,"[52] but since "the rules of hygiene demand the disappearance of all things corrupt as soon as possible," Fabre praises dung beetles and scavenger species for their "services rendered" as "sanitizers" and as "fertilizer buriers."[53] The metaphor of the profession also reflects a series of gestures, particularly noticeable in the activities of construction or building of a cottage or in the preparation of food reserves. A good example of this last activity is offered by the minotaur beetle, whose female is the "baker," while the male "brings her what she needs to make flour." The division of labor inspired this comment from Fabre: "As in any good household, the mother is the minister of the interior, the father is that of the exterior."[54] House building was fairly well represented by a cricket digging a burrow: "The miner scrapes with his fore-legs, and uses the pincers of his mandibles to pull out the larger bits of gravel. I see him stamping with his powerful hind-legs, furnished with a double row of spikes; I see him raking the rubbish, sweeping it backwards and spreading it slantwise. There you have the whole process."[55]

A miner digs his house, while the mason builds it. In this case, Fabre's example is female, as he is describing mason bees. Fabre appreciates the common name, "which paints them with a word," but still gives the scientific names of the two species he described: *Chalicodoma muraria* and *Chalicodoma sicula*. Sometimes two techniques serve the same purpose. Fabre does not hesitate to spin the metaphor by presenting two Hymenoptera, the bembex and the tachyte, as "two workers of the same profession" who employ "different methods . . . to arrive at the same result," and he concludes that "each of the builders has his own particular art for his estimates, his clientele."[56]

At first glance, using professional metaphors could appear to push popularizing rhetoric beyond reasonable limits. However, Fabre's all-out use of anthropomorphic metaphors, which creates scenes and invites readers to discover specific facts and gestures, illuminates insect behavior more than Fabre's dogmatic conception of insects. This metaphorical language corresponds to an essentially descriptive and narrative mode of exposure that reinforced the feeling of reality.

WRITING LIKE AN ENTOMOLOGIST

In his course on literary theory, Antoine Compagnon, a professor of French literature at the Collège de France, demonstrates just how far one can stretch the notion of literary genre, which he considers not as a rigid category but as a hypothesis that a reader develops about what he expects to find when he prepares to read a book.[57] The flipside of this definition is that when the reader finds herself in a text written in a characteristic style, she recognizes the text as belonging to such and such a literary genre.

In that respect, Fabre's opus, taken as an archetype of entomological description, borrows from several literary genres, including adventure novels, fantasies of devouring desire, comedies of manners reminiscent of Balzac or Pagnol, intimist novels, and even two fables in Provençal. A writer reading all of Fabre's *Souvenirs entomologiques* will find a great number of familiar literary forms.

To say an author observes her contemporaries like an entomologist not only suggests that she observes them as she would insects, but it also recognizes that the manner she uses to describe them stems from rhetoric and literary forms that Fabre and other entomologists borrowed from literature, which would be to say that she writes like an entomologist, with literary expression that runs the gamut from horror and disgust to interested astonishment and fascination, along with humor and even tenderness.

The presence of entomology on the literary scene remains marked by ambivalence, even in Japan where *Souvenirs entomologiques* was translated three times, including once at the beginning of the twentieth century by the militant anarchist Sakae Ossugi (who also translated *On the Origin of Species*). Exhibits and books have made Fabre a familiar figure, all of which goes hand in hand with insects being quite present in Japanese culture, even as early as the Edo period, when the Japanese painter Kitagawa Utamaro produced his illustrated books of insects.[58] Younger generations have not escaped this enthusiasm, with many children and teenagers building cages to raise insects and participate in entomology workshops.[59] That said, discomfort with insects, and even anxiety over them, also exists, as seen in a schoolgirl's comment that she is afraid of moths because of their unforeseeable flight. The famous novel by Kobo Abe *The Woman in the Dunes,* made into a movie, presents a tragic dimension, staging an entomologist trapped in a funnel of sand, which is reminiscent of how the larvae of certain species of four-milion capture insects, often ants, which they prey on. Was Kobo Abe, author of the book and the screenplay directed by Hiroshi Teshigahara,[60] trying to express the potential anxiety caused by insects or by this contrary image of a trapper being trapped, paying homage to the insect victims of human curiosity? Whatever the case, Kobo Abe demonstrates enlightened knowledge of entomology, using precise names of species and describing their behavior and habitat correctly.

An entomologist's writing, like the writing of authors who have studied insects, is filled with scientific names, which are for the most part unknown by nonspecialists. In Europe, insects are generally so unknown that most of the time they do not even have common names that allow one to distinguish different species. They are named by the family, in a taxidermic sense of the word, to which they belong—flies, ants—as if we only had the word *felines* to differentiate tigers from lions and cats. An entomologist's writing can also be seen in the precision of the vocabulary used. The style can be light without it damaging the scientific seriousness of the text.

PERSPECTIVES ON ENTOMOLOGISTS

While entomologists convey their observations about insects and other arthropods with a characteristic diversity of styles combined with descriptive and taxonomic precision, their observations are characterized by gestures of capture, ensuring the choice of places to observe the creatures, a practice that can be found to a certain degree among enlightened amateurs and entomology lovers.[61] As a result, one is tempted to look more closely *at* entomologists.

Entomology is a profession, with its degrees, field research and laboratories, and results published in peer-reviewed journals.[62] At the same time, there exist a whole range of amateurs: dilettantes, occasional local researchers, volunteers spending their free time studying entomology and occasionally doing research that is useful to professionals, along with self-taught specialists who are eventually specialists in another domain.[63]

In any case, be they amateurs, dilettantes, volunteers, self-taught practitioners, or professionals, entomologists are usually—or at least were for a long time—male, and rarely female. Women would receive lessons, like the Julie to whom her husband Étienne Mulsant addressed an initiation in the form of letters.[64] However, talent as an illustrator allowed women to access work on insects. Madeleine Pinault-Sørensen establishes this point in *Le Peintre et l'histoire naturelle*.[65] The paintings of insects made in Surinam by Maria Sybilla Merian give her a prominent place in the history of entomology, the precision of her observations being intertwined with aesthetic quality. We should also mention Hélène du Moutier de Marsilly, a collaborator and friend of Réaumur, whose meticulous drawings help clarify the anatomy of bees, among other things.[66] Christine Jurine, daughter of the Genevan naturalist and doctor Louis Jurine, contributed to her father's entomology work—especially on Hymenoptera wing shapes—and drew illustrations.[67]

Generally speaking, by engaging in their hobby, amateur entomologists of past centuries helped to collect, describe, name, and classify a multitude of insects and other arthropods.[68]

Professional entomologists would not have been able to undertake this inventory without help. Today, amateurs still play a key role. In the field, passionate amateurs can always be found side by side with retired academics and motivated students.[69] The term *participatory science* has been coined to account for this kind of amateur involvement, and it can be evidenced by the success of the Garden Butterfly Observatory launched by the French National Museum of Natural History.[70] Does this success mark the beginning of a change in European culture with regard to insects? It could be interpreted as entomology being taken seriously, with the folk image of the butterfly hunter being replaced by that of a researcher measuring the effects of human activities on the environment, in which insects play a vital role. This change of perspective regarding entomologists, whose role as experts is taking on a political dimension, with the continued help of knowledgeable and motivated amateurs, also marks an awareness of the need to think about the complex relationships between science and politics. These relations, which now face the challenges of the decision-making game, have long existed in a more metaphorical form, where the political dimension of science also found places where it was expressed and debated.

4

INSECT POLITICS

The saga of honey—this is one way to summarize the pages that Virgil devotes to beekeeping. The poet is addressing Maecenas, who ordered the Georgics, and promises: "I'll speak about the celestial gift of honey from the air." He adds: "I'll tell you in proper sequence about the greatest spectacle of the slightest things, and of brave generals, and a whole nation's customs and efforts, tribes and battles."[1]

Two digressions followed. One, rather brief, focuses on the plants surrounding hives and relates to a small garden arranged by an old man at the base of the walls of Taranto. The other, more developed, tells how Aristeus loses his bees because he involuntarily causes the death of Eurydice and, consequently, that of Orpheus, and how he finds a new swarm in the bowels of four sacrificed bulls. The first digression evokes an economic and aesthetic optimization of space, while the second, from our perspective, stems from legend and ritual. Yet we are the ones who retrospectively establish this division. Whether viewed from a technical or magical point of view, whether efficacious or illusory, this text refers to apicultural know-how that for Virgil, as for a large number of authors, coexists with descriptions of bee behavior using political metaphors.[2]

Dating three centuries earlier, the Septuagint, the Greek translation of the Old Testament, contains another eulogy of honey and the insects that produce it. One of the books, Proverbs, advises those who are lazy to follow an ant's example. Following this injunction comes one to observe toiling bees, whose honey is sought after by kings and commoners, and who, despite their small size, should be admired for their wisdom.[3]

KING OR QUEEN?

When observing a beehive attentively, one of the first questions that comes to mind is that of how the labor is divided.[4] Without counting the larvae, one can distinguish three types of individuals that have different morphology and even more different behavior. The most numerous, which can number tens of thousands, have a stinger and pollen sacs on their back legs. They build the honeycombs, take care of the larvae, collect nectar, gather pollen, and defend the colony against intruders. The others, less numerous and a bit larger, are called drones and seem to live at the expense of the former, which kick them out before winter. Finally, there is a single bee, larger and longer than the others, which seems to play a regulatory role and upon which the image of royalty is conferred. Determining the male or female sex of this metaphorical sovereign figure is one of the key questions related to representing the social life of bees.

Apparently, ancient authors imagined hives to be ruled by a king and not a queen. At least, that is what we are tempted to believe, since this representation corresponds to a traditional distribution of social roles between men and women.

However, *Oeconomicus* by Xenophon, which was written around 370 BCE, calls for a more nuanced viewpoint. Xenophon is a Greek aristocrat, warrior, and writer, who is also interested in agriculture. Having studied under Socrates, he imagines the latter dialoguing with a friend, discussing the explanations a landowner would give to his wife regarding household management. A woman's tasks are defined based on an analogy with the work that Xenophon considers to be that of the "queen bee in the hive."[5] She "sends forth [the bees] to their labours; and all that each of them brings in, she notes and receives and stores against the day of need," she presides "over the fabric of choicely-woven cells" and "under her guardian eye the brood of young is nursed and reared." When required, she encourages them to depart and look for a new leader. When the wife, who is clearly a bit perplexed, asks if that is the task she is expected to complete, the husband responds that indeed she is to send out the servants whose job was outside and that she is to ensure that those who stay in the home are doing what they should be doing, and that, in addition, she should receive and distribute food and other resources. These all being things the wife is already doing, they serve more as a description than a prescription, and the reference to the queen bee is meant to legitimate her work. This comparison is not obvious. In the analogy between the hive and the family estate, the mistress of the household has purely domestic power, which for Xenophon only has meaning as a complement to political power, which is linked to the strictly masculine domain of being a soldier. Although Xenophon does not explicitly express it, we can deduce from this text that the queen bee's power is exclusively feminine because there are no wars among the bees, or, more precisely, because Xenophon does not speak of war among bees.

The link between the sovereign's gender and the state of war underlies Aristotle's analyses.[6] In *the History of Animals,* Aristotle bases his arguments on the distinction between "bees," which have a stinger, "drones" that don't, and "rulers," which have a stinger but don't sting.[7] He goes further in the treatise called *Generation of Animals*, in which he suggests a series of possibilities. To understand the meaning, it is important to keep in mind that he uses the word *bees* for what we call *worker bees*. Aristotle formulates the following alternative: either the "bees" are born from the union of "bees," "drones" from "drones," and "kings" from "kings," or all are generated by one kind, or by the union of two kinds, such as drones and bees, and, in that case, either the drones are male and the bees female or the contrary.[8] Yet, for Aristotle, the hypothesis that drones are male and bees female is not reasonable, "because Nature does not assign defensive weapons to any female creature," and bees have a stinger and drones do not. Mixed in with reasoning full of detailed observations, we find an accepted notion of the time regarding the social division of the sexes serving as an argument.

A similar alliance between empiric knowledge and social representations can be found in the first century of our era, in the pages of Pliny the Elder's *Natural History*, book II, chapters 4 to 23:

> But among them all, the first rank, and our especial admiration, ought, in justice, to be accorded to bees, which alone, of all the insects, have been created for the benefit of man. They extract honey and collect it, a juicy substance remarkable for its extreme sweetness, lightness, and wholesomeness. They form their combs and collect wax, an article that is useful for a thousand purposes of life; they are patient of fatigue, toil at their labours, form themselves into political communities, hold councils together in private, elect chiefs in common, and, a thing that is the most remarkable of all, have their own code of morals.[9]

Products of the hive—honey and morals—are two themes that constantly repeat themselves in literature about bees. As for the still obscure question of the ruler's sex, Pliny speaks of a king, in the masculine, as if that were to be expected.

> The ruling bee never does use a sting. The obedience which his subjects manifest in his presence is quite surprising. When he goes forth, the whole swarm attends him, throngs about him, surrounds him, protects him, and will not allow him to be seen. At other times, when the swarm is at work within, the king is seen to visit the works, and appears to be giving his encouragement, being himself the only one that is exempt from work.[10]

AMAZONS AND THE MICROSCOPE

It took sixteen centuries for the masculinity of the sovereign to be called into question. The key book on the topic is Charles Butler's *The Feminine Monarchie or the Historie of Bees*, which was first published in 1609. There were subsequently several reprints through to the eighteenth century. The author's technical know-how gives his work the air of a beekeeping manual, which explains its popularity, yet Butler has no fear of entering the realm of politics. The hive is, in his words, "an Amazonian or feminine kingdome."[11] He comments that the Drone "is but an idle companion, living by the sweat of others' brows."[12] Bees (that is, worker bees) rightly manage and dominate the drones. In a word, the hive can be described using the words of a grammarian: "the feminine gender outweighs the masculine." Worried that a reader could draw "sophist" conclusions regarding the possible status of the two sexes among humans, Butler is quick to specify that bees are a special case and that, generally, males dominate females.[13]

The question of the leader's male or female gender is part of a larger issue regarding determination of sex among the various bee forms found in a hive, the majority of which are sterile females, considered by some authors, following Réaumur, as hybrids or "mules." We already mentioned ants regarding their supposed hard-working nature and their legendary supplies.[14] Cited as an example to follow, ants are also perceived as inconvenient and worrisome due to their numbers. When Montaigne deplored that "there is more ado to interpret interpretations than to interpret things," he summarizes his thought with the phrase: "Tout fourmille de commentaires."[15] He uses the verb *fourmiller*, which stems from the word *fourmi* or ant. This idea of proliferation translates into English with the image of a swarm, and Montaigne's phrase is rendered as "Everything swarms with commentaries." Swarm refers to large, dense groups of flying insects, behavior attributed to bees. And one of the most astonishing aspects of ant life is without a doubt the seasonal appearance of winged ants. Although many city dwellers today believe they are a specific species of ants, naturalists have long observed that flying ants come from anthills. Some even think the wings were some form of gift nature gives to ants that were on the verge of death.[16] It is difficult to determine how woodcutters, shepherds, and others who often had the opportunity to observe an anthill in the past perceived flying ants. We do nonetheless find a precious indication in *Don Quixote*. Cervantes has Sancho Panza pronouncing the proverb "To her hurt the ant got wings," and when he abandons his supposed governor's position, Sancho declares: "Here in this stable I leave the ant's wings that lifted me up into the air for the swifts and other birds to eat me."[17]

The questions whether the hive is led by a king or queen and why some ants have wings at certain times would both be answered during the same period, thanks to anatomical observation.[18] The decisive turning point occurs during the second half of the seventeenth century, in Dutch naturalist Jan Swammerdam's 1669 publication *Historia Insectorum Generalis* (The natural history of insects). He sets apart "bees that make honey"—in the original Latin, *operariae*, or workers—which he states "in which we have discovered no parts that would allow us to conclude that they are male or female."[19] On the contrary, he adds, "in the kings which are the drones, and the queen (mistakenly called the king), the parts that serve reproduction are quite perceptible."[20] In other words, by dissecting and observing the insects under a microscope, Swammerdan is able to recognize the genitals and establish that drones have male characteristics and that the sovereign is female. The king is in fact a queen and a mother. On the other hand, the sex of the other bees remains undetermined. Regarding ants, Swammerdam deduces from his observations that large winged ants are male, and he considers the mass of wingless ants to be "workers."[21] Swammerdam's conclusions are summarized in a few lines: "Male ants therefore only dominate among the ant colony when they go about propagating the species, and that is the same among bees, with whom ants share much in common. At all other times, these two kinds of insects, whose instinct leads them to live in troops, know no subordination, and form kinds of republics where everything is shared and where all are equal."[22]

The strangeness of insect mores, and, in particular, the reversing of traditional gender roles among bees, is mentioned by the French writer Bernard Le Bouyer de Fontenelle in his 1686 *Entretiens sur la pluralité des mondes* (Conversation on the plurality of worlds).[23] This work combines elegant expression and rigorous knowledge. Fontenelle recounts an evening in a park as he is trying to seduce a curious marquise by talking to her about physics and astronomy. Toward the end of the third evening, asked to give a more precise picture of the beings that inhabit various planets, the narrator assures his companion that there is a planet, which he does want to name, that has "very lively, hard-working and skilled inhabitants." Although we could hold it against them for being looters, we had to admire "their zeal for the good of the State." Sterile, they are forced to be chaste; their "nation" is nevertheless perpetuated by a "queen" of "astonishing fecundity" and that has, for its pleasures and propagation, a seraglio of husbands, whose sole function is to impregnate her and who are killed once they have fulfilled their office. When the marquise contests this "fiction," the narrator then reveals that he is talking about bees. He uses the analogy here as a way of introducing scholarly controversies to a worldly crowd.

On the other hand, we find no trace of these preoccupations in Bernard Mandeville's *The Fable of Bees*. This classic in political literature, published in 1714, borrows from hives nothing more than the image of a wealthy, strong, and populated city. The author imagines that bees grow tired of their prosperity based on avidity and fraud and decide to become virtuous. The immediate result is to ruin lawyers, jailers, law keepers, who are no longer needed as there is no more crime. There is no more use for locksmiths either, as there is no more theft. Doctors prescribe local remedies that are more useful and less costly than exotic remedies. Holders of religious offices are keen to do their duty. The richest bees leave. The price of land and homes plummets. People working in luxury fields, including painters and sculptors, end up unemployed. The deserted hive is attacked by neighbors, and the last remaining bees, vanquished despite their bravery, find themselves holed up in a hollow tree trunk. This fictional account, whose moral is summarized by its subtitle "Private Vices Publick Benefits" is half a century ahead of the "invisible hand," which, according to Adam Smith, leads individual to contribute to "society's annual income" by pursuing their own private interests.[24]

Quite different from the *Fable of Bees*, the *Spectacle de la nature* (The spectacle of nature), by the Abbey Noël-Antoine Pluche, was a popular book in libraries in the eighteenth century.[25] It combines popularized science and religious propaganda. When speaking about bees, the author expresses his endless admiration for "these little animals" and their "spirit of society." He highlights that they "are free because they only depend on laws" and they are rich "because the complementarity of their various services" ensures sufficient abundance. By comparison, "human societies" appear "monstrous": "to become superfluous, half of humanity removes from the other half their basic necessities."[26] Behind this political commentary is a theological lesson: "As long as men are not led by the spirit of God they are quite easily the most unfair and corrupt of all animals."[27] The author's pessimism leads to the conclusion that without God, men are inferior to animals—and this pessimist implicitly agrees with the Jansenist vision of man's misery in the absence of grace. The comparison is far from fortuitous. Michaud's biography of Abbey Pluche in *Biographie universelle* informs us that he was reprimanded by his superiors for opinions contrary to the *Unigenitus* bull, which means that his superiors did indeed suspect him of having Jansenist sympathies. Despite all that opposes them, Mandeville's bees and Pluche's bees share a vocation of political commentary.

COMPETING PARADIGMS

The observation of social behavior and the anatomical determination of polymorphism form a conceptual foundation, one open to additions and corrections. The notion of a paradigm as defined by Thomas Kuhn could apply to Swammerdam's work, but particularly to Réaumur, whom we have already cited for his relative disinterest regarding issues of classification and nomenclature as well as for his controversy with Buffon regarding the role insects play in natural history. The *Mémoires pour servir à l'histoire des insectes*, published in six volumes from 1734 to 1742, were illustrated with 267 plates whose quality highlights the importance of anatomy, which is capable of explaining behaviors.[28] Many pages in these six volumes are dedicated to bees. As we already mentioned, a text about ants written for a seventh volume remained unpublished for a long time.[29]

In entomological memory, bees and ants are linked to the names of two Swiss authors who marked the discipline: François Huber and his son Pierre Huber.[30] François Huber's *Nouvelles observations sur les abeilles* (Observations on the natural history of bees), first published in 1792 with a second edition in 1814, is considered a seminal work throughout the nineteenth century, notably due to observations regarding fertilization of queen bees. The work combines a series of letters addressed to Charles Bonnet. The author went blind at the age of twenty, yet he makes numerous observations and experiments on bees with the help of his servant, François Burnens.

Pierre Huber worked with his father and also carried out his own research into ants.[31] Published in 1810, his book entitled *Recherches sur les moeurs des fourmis indigènes* (and translated into English in 1820 under the title *The Natural History of Ants*) has still been cited in recent works, notably with regard to his observations on the role of antennas in ant communication and on the relationship between aphids and ants, but mostly with regard to the discovery of a form of social parasitism, which he refers to as slavery, a term that has caused a certain amount of controversy.[32]

The Hubers opened a field of study in entomology—that of social insects. This denomination is not in itself obvious. To speak of social insects, one needs to define a group of species solely by their social behavior, while it is more common for morphological characteristics to carry more weight in classification systems.

In this regard, Lepeletier's work plays a unique role in entomological history. Amédée Louis Michel Lepeletier, count of Saint Fargeau, came from a family of nobles of the gown and was the youngest of three siblings, his two elder brothers being active in the French Revolution.[33] His main work, *Histoire naturelle des insectes: Hyménoptères* (Natural history of insects: Hymenoptera), was published in 1836. For him, the difference in the "instinctive faculties" is a "distinctive characteristic of families" that is more important than anatomical difference "that are simply an expression of the former." Furthermore, he gives this audacious methodology a metaphysical justification: "It would seem that the Author of creation gave animals instinct as reason above matter; and as a result, it is a law to consider the intelligent part as serving as a mold for the body."[34]

The path set out by Lepeletier was not followed and, undoubtedly, was more appealing than it was useful. Characterization of insects by their social behavior does not coincide with their position in the classification. Since the beginning of the eighteenth century, ants, bees, and wasps have been classified as Hymenoptera, an order in which social species are a minority. Termites, which were long unknown to Europeans (or confused with ants and even called "white ants"),[35] were included among Neuroptera at the beginning of the nineteenth century and were then considered to form an order on their own, the Isoptera, that now includes twenty-eight hundred species. Furthermore, although thousands of ant species live as social groups, many bee species have solitary behavior.[36]

That said, the exclusion of behavioral characteristics is less radical than it appears. Indeed, it remains possible and useful to relate behavior and morphology. Acoustic communication, for example, presupposes organs that are adapted to the emission and reception of sound signals.[37]

Such competing paradigms are a sign of the abundant research and observations that occurred during the eighteenth century and the first half of the nineteenth century, some of which remained marginal while others became the foundational bricks of entomology. From Swammerdam to Huber, passing by Réaumur and Latreille, the study of social insects became a structured field while at the same time political-sounding controversies took the front stage.[38]

REPUBLIC OR MONARCHY?

When Jules Michelet publishes *L'Insecte* in 1858, he has a large quantity of high-quality documentation.[39] He voluntarily admits that his knowledge comes from his reading, and the historian insists on the "a strong and decisive blow" his mind received "from the books of the two Hubers on the Bees and the Ants."[40] It wasn't until later, under the influence of his second wife Athénaïs Mialaret and with her help, that he undertook to write about natural history.[41] In *L'Insecte*, and in his three other popularized books on natural history, *L'Oiseau* (1856), *La Mer* (1861), and *La Montagne* (1868), Michelet uses a naturalistic descriptive approach in order to develop an often metaphoric spiritualist vision of nature.[42] In his introduction, he affirms that the insect world appears to be "a world of mysteries and gloom" in which "nevertheless, the most penetrating light is thrown on the two cherished treasures of the soul—Immortality and Love."[43]

Twenty years after the publication of *L'Insecte,* the book is recognized as a historic stepping-stone that has since become outdated. The French thinker Alfred Espinas, who published *Les
Sociétés animales* in 1877,[44] praises Michelet "This great historian
spoke of the animal family as nobody had done before,"[45] but his
sources are different. He does occasionally cite François Huber
and, more frequently, Pierre Huber, yet Espinas's main reference
is Auguste Forel's *Les Fourmis de la Suisse* (Ants of Switezerland),
which was first published in 1874.[46] For Espinas, insect societies
are but one of the biological manifestations of the social. They
are simply "maternal domestic societies" and are superior to societies that rely only on nutrition and those that bring together
individuals only for coupling, but they remain inferior to "pater-
nal domestic societies," which can be observed among birds and
mammals.

Forel and Espinas opened an era that leads right up to William Morton Wheeler's courses given in Paris in 1926. For the
French speaking, it is dominated by the flamboyant trilogy by
the Belgian essayist Maurice Maeterlinck, made up of *La Vie des
abeilles* (The life of the bee, 1901), *La Vie des termites* (The life of
the termite, 1926), and *La Vie des fourmis* (The life of the ant,
1930).[47] The novelty lies in an affirmation of the evolutionary perspective in the study of behavior. Despite this upheaval, it is still
possible to follow certain recurring themes over time.

The term *republic* is found frequently in descriptions of beehives and ant colonies, yet for some authors is not incompatible
with the kingdom metaphor. An article titled "Abeille" (Bee) in
a *Manuel du naturaliste* (Naturalist's guidebook), whose second
edition was published during year 5 (1797), states that the attach-
ment bees have "for their queen is equal to how useful she is for
the republic."[48] Here the republic seems to be a synonym for state.
This original meaning refers back to the Latin *res publica* and

justifies the use of the word *Republic* to translate Plato's work, which in Greek was called *Politeia*. This is the meaning that Latreille uses in 1802 when describing ant colonies as republics sheltered from agitation. In 1810 Pierre Huber calls the last chapter of his *Natural History of Ants*, "Observations on Those Insects That Live in Republics," referring to wasps as well as bees, ants, and termites.[49]

However, although the word *republic* is used frequently, for certain authors it continues to carry tones of antimonarchical controversy. In 1822 the author of a beekeeping memoir championed the bee queen: "The mother bee leads a monarchy and not a republic, despite what has often been said. And, what a monarchy it is! What wisdom, what love of the public good can be found in the leader's laws. And on the other side, what devotion, what patriotism, what union can be found in the subject. May it please God to see monarchies in Europe offering the same spectacle."[50]

Lepeletier de Saint-Fargeau holds the idea of monarchy in high esteem when he sets out to demonstrate that there are no queens among ants: "Watch over the interests and needs of other members of society, provide useful orders, these are the duties of royalty; being obeyed is its right. All that we have seen in ant colonies so far is removed from the idea of orders given, and although everything is done in agreement and with punctuality, it is not because one single mind developed the idea to be executed."[51]

So we have monarchical bees and republican ants. These ideas are congruent with Michelet, who affirms, "The ant is frankly and strongly republican," and that, unlike the bee, it does not even need to find "a moral support in the worship of the common mother."[52] Not all authors are duped by these images. After using them, Pierre Huber warns the reader: "The terms of

Queens, of Subjects, of Constitution, of Republics, must not be taken according to the strict letter."[53] In 1795, or year 3, French naturalist Louis Jean-Marie Daubenton gives a course at the École Normale criticizing the errors induced by the "pompous style" used in natural history: "The lion is not the king of animals," he states, directly implicating Buffon. In front of an enthusiastic audience, he adds "there is no king in nature." When the students are then invited to ask questions, one objects: "I saw something worse than a king in nature, that is, I saw a queen and, what was even more extraordinary was that it was a queen in a republic." Daubenton responds that the power in the hive is held by the "worker bees" and that the so-called queen is nothing but a mother who seems to have no other role than spawning.[54]

INEQUALITY AMONG INSECTS

Insect societies may be republics, but they are far from being egalitarian. First of all, among bees and ants equal balance between the sexes is far from being. Males only serve for fertilization. In 1798 Latreille notes that after the nuptial flight they do not return to the ant colony, where their presence would no longer be useful. "They fulfilled the desires of nature, those of love, and are already no longer; the reign of pleasures is so very short."[55] Pierre Huber brings up queen bee fertilization when he mentions a "male seraglio."[56] Lepeletier de Saint-Fargeau defines touch as "that sensation so necessary for reproduction of the species," but does not seem at all moved by the massacre of drones by worker bees.[57] The crucial point is, clearly, the reversal of traditional gender roles. As Pierre Huber writes, "But how strikingly does it contrast with our manners, that the arms, courage, military skill, should, in these republics, rest with the female sex; whilst feebleness, idleness, and exile, fall to the lot of the males."[58]

Termite societies do not demonstrate this gender inequality, but they are not egalitarian. The poet Jacques Delille describes them as follows:

> In three classes is organized their wise republic,
> Happy peoples of worker, nobles and soldiers.[59]

Delille undoubtedly places the reproducing couple in the class of "nobles." This biological division into three casts evokes the medieval three-part division of roles (peasants, knights, and clerks), and yet one would hard-pressed to find the equivalent of a clergy in insect societies.[60]

Among ants and bees, cast inequality accompanies gender inequality. This is particularly distinct in ants, for whom nature forbids workers the "sweet pleasures" of love, while also depriving them of wings, condemning them to be "helots" forever attached to the ground.[61] But Latreille does not limit himself to this pessimistic vision and suggested a reversal:

> Authority, power, and strength reside primarily in these small beings that appear to us to be so disgraced. They are the nurturers, the tutors of a family. The existence of the numerous posterity is confided to their care. The education of the adopted children is undoubtedly a source of great happiness to that, and this participation in the maternal role provides them pleasures that make up for those of which they are deprived.[62]

Other authors share Latreille's argument. Pierre Huber marvels at the bond nature inspires in workers, attaching them "so strongly to the progeny of another mother."[63] Latreille and Huber recall that workers can even elect a new queen.[64] In any case, the workers are the game masters, they are the ones who collectively wield maternal authority over the hive and ant colonies. Michelet develops these themes lyrically, entitling his final chapter "How the Bees Create the People and the Common Mother."

These matriarchal republics find a zealous supporter in the person of Alphonse Toussenel, a naturalist and popularizer, a disciple to Charles Fournier, and an author Léon Poliakov describes as an anti-Semitic firebrand.[65] Toussenel is convinced of the superiority of the feminine nature, and he views bees and ants as providing a demonstration that "individual happiness" is "directly due to feminine authority." He presents the hive as the only republic in which wealth depends on work alone. To counter possible objections to the drone killings, he argues "that it is difficult to make an excellent omelet without breaking some eggs."[66]

This harmonious image of insect societies, which apparently could not be questioned despite the unfortunate destiny of the males and the forced chastity of the workers, will nevertheless be disrupted by war and slavery.[67]

ON WAR AND SLAVERY

First war. Pierre Huber recalls Virgil's bee wars and sees in ant wars "a strong image of our serious quarrels."[68] Lepeletier de Saint-Fargeau, who justifies the killing of the males by their incapacity to work, judges more severely the "combats between worker bees from different hives." He finds "little or no excuse" for them and determines that their causes lay in the nature of bees: "neither hospitable, nor generous, they are

sometimes thieves."[69] Michelet would dedicate an entire chapter to describing, with horrified fascination, what he calls "civil
war,"[70] incorrectly for that matter, because the war occurs
between ant colonies of different species, the tiny masons wear-
ing themselves down taking revenge on the large attacking car-
penter ants.[71]

Slavery sparks more emotion and reflection than war.[72] This
begins with Pierre Huber himself, who recounts his discovery
of it, recalling the moment very precisely: "On the 17th June,
1804, whilst walking in the environs of Geneva, between four
and five in the evening, I observed close at my feet, traversing
the road, a legion of Rufescent Ants."[73]

He follows them and sees them attack a nest of dark ash-
colored ants, removing the larvae and pupae and carrying them
back to their own nest. He decides "to give them the appellation
of Amazon or Legionary Ants." Thus he discovers "mixed or
compound anthills" in which the captured ants serve the amazons as workers. To explain this phenomenon, he goes so far as
to use the analogy of slavery: "The Ash-coloured and Mining
Ants, are to be considered then as the negroes of the Amazons;
it is from among them the latter procure slaves; they kidnap them
at an age when their instinct is not developed; and these insects,
on being brought up by the Amazons, divide with them the fruit
of their industry."[74]

This situation does not bother him from a moral point of view.
On the contrary, he admires the prudence with which nature
establishes among insects this institution "somewhat barbaric
among men." In mixed anthills, one sees neither "servitude" nor
"oppression," and the slave ants do not seem to "entertain the
slightest suspicion of their being in a foreign nest."[75] Based on
the way Pierre Huber talks about slavery among ants leads one

to believe that Huber is shocked by the acts of cruelty found in human slavery but does not question the institution itself.

The French naturalist and anthropologist Julien-Joseph Virey will seek to demonstrate that "although nature seemed to have created slavery among ant species," in fact, these slaves become citizens of their new state. The issue of servitude is thus reduced to an apparent inequality among castes, and to the lot of helots: "Each of the castes having a determined job and carrying it out dutifully, one could say that the superior ants are neither more free, nor more in control than the subordinate ants." And he contrasted this happy state of affairs to that existing among men, where some "have subjugated and enslaved others."[76]

Lepeletier de Saint-Fargeau is more cynical and views the slavery as a manifestation of a comparative mind found in animals that are deprived of reason. He admires an insect that "feeling the sensual delight it will have at rest, attains it by procuring loving servants that do all the work of domestic service sparing it the fatigue."[77] He emphasizes that there was no risk of a new Spartacus among the slaves, as they were captured so young they "know only the homeland they serve."[78] The same antiquating rhetoric can be found in the writing of Jean-Louis Armand de Quatrefages, in his *Souvenirs d'un Naturaliste* (Memoirs of a naturalist), in which he describes ants that "lead the leisurely life of formerly warring people," knowing like the latter how to be "served by slaves."[79]

Michelet reacts differently. He describes how he discovers ant slavery by reading Pierre Huber.[80] More than astonished, he is indignant. How could a man who turned to nature to find innocence find something "without a name!"[81] He imagines triumph for the partisans of slavery. In his anger, he attacks Huber's book. And then he decides to examine the facts more closely, carefully differentiating the horrors of slavery from bucolic farming. There is nothing more normal than ants maintaining aphids and licking their honeydew. As he says: they "are cattle, and not slaves."[82] He then moves on to slavery among ants, recounting in turn Pierre Huber's discovery, describing the mixed anthill, marveling over these "civilized helots" that "love their great barbarous warriors" and care for their children. Finally, he tries to explain the slavery by division of labor. All ant hills have at least three casts: male, fertile female, and the mass of "laborious virgins." This mass, which Michelet considers to be "truly the people" is divided into a class of warriors and one of workers. Thus he suggests slaving species are monstrous forms that, following some migration, are missing the essential class of industrious ants: "Therefore, that they may not perish, they carry off the little black souls, which tend them, it is true, but also govern them."[83]

Thus nature is far from justifying slavery, but rather denounces it as monstrous and in a form of justice makes the masters dependent upon the slaves. Michelet has turned away from history toward nature, only to find history in nature. Here we join Roland Barthes's judgment: "Michelet did not naturalize morality, he moralized nature."[84] Among the authors who naturalize morality regarding social insects, one finds an illustration in Marcelin Berthelot, a famed politician and chemist who observes insect in his spare time. In 1886 he publishes a collection of essays called *Science et philosophie* (Science and philosophy), one of which is called "Animal Cities and Their Evolution."[85] According to Berthelot, comparing human and animal societies reveals a similar instinct of sociability, which renders the social much better than the "fanciful" social contract hypothesis. Berthelot, familiar with the typical objection that the stability of animal societies contrasts with the changes that affect human societies, opposes this argument, citing the example of the vicissitudes of an ant-hill he observed in a forest near Paris. Ten or so years later, in a collection of articles called *Science and Morality*, Berthelot dedicated an essay to ant invasions.[86] He specifies that it is more useful to compare human societies to anthills than to beehives because in the latter the laws are uniform, while in ant societies there is room for individual initiative. In 1903 Marcelin Berthelot contributes to a new edition of *L'Insecte*, providing an opportunity for him to pay homage to Michelet as well as to criticize the manner in which he sought out the symbolism of his own thought in natural history.

However, Michelet is not the only author to whom Barthes's phrase could apply. Darwin himself does not entirely escape the temptation of moralizing nature. In chapter 7 of the first edition of *On the Origin of Species* (1859)[87], he refers to Pierre Huber's work when discussing the slave-making instinct in some ants, and he imagines that the ancestors of slave-making ants have limited themselves to stealing and storing larvae from other species as a source of food. Some larvae would develop and turn into workers, which would then follow their instinct and do the work they knew how to do in the new anthill. For Darwin, the genesis of behavior could be explained by the same type of reasoning as the formation of an organ. The reference to morality comes in the form of an incidental remark. Indeed, Darwin explains that he wants to observe what Pierre Huber observed himself, because he cannot help but doubt the existence of "so extraordinary and odious an instinct as that of making slaves."[88] This sentence is surprising. Why would a naturally occurring odious fact be more difficult to admit, unless one applies some sort of moral value to nature?

EVOLUTION AND SOCIETIES

A moral vision of nature also underlies a number of Jean-Henri Fabre's interpretation. We see this illustrated in the pages of *Souvenirs entomologiques* devoted to "pine processionaries."[89]

Fabre recounts their summer egg laying and hatching several weeks later. He describes the caterpillars building nests with the first cold weather, collective homes in which they would spend winter. He shows how they move in long columns on trees they could threaten if their numbers are too great. Finally, he watches the formation of cocoons and the hatching of the moths, called pine-tree lappets, which copulate and then finish their cycle by laying eggs that will hatch into a new generation of caterpillars. Nest-building retains his attention: "High up in the pine the tip of a bough is chosen, with suitably close-packed and convergent leaves The spinners surround it with a spreading network, which bends the adjacent leaves a little nearer and ends by incorporating them into the fabric. In this way they obtain an enclosure half silk, half leaves, capable of withstanding the inclemencies of the weather."[90]

One could think that a building built with such care would be defended staunchly by its first occupants. This is not the case, and Fabre is able to remove a group of caterpillars from one branch and transfer them into an already occupied nest, without the new occupants being accosted by the current occupants. Words such as "Everything for everyone" and "Each for all and all for each" seem to guide the social life of these caterpillars, who "know nothing of property, the mother of strife" and practice the "strictest communism." So the study of social behavior among the larvae of a Lepidoptera leads to a digression into political philosophy. From the start, Fabre adopts an empirical position. He hears "generous minds, richer in illusions than in logic, set communism before us as the sovereign cure for human ills." Yet, says Fabre, communism is possible among pine processionaries because they do not face the "food problem," as a pine needle is a sufficient meal for a caterpillar, and they know "as little of family ties."[91] Maternity requires a mother to think first of all about her progenitors. Without property, family, and sexuality, pine processionaries achieve, in Fabre's eyes, a uniformity and a leveling that, he concludes, would be both impossible and unwanted if applied to man.

Through their observations of the insect world, Fabre and Darwin reveal their political inclinations. However, what separates them is decisive. It is true that the two authors appreciated each other. From the first edition of *On the Origin of Species*, Darwin confidently refers to an observation made by his French colleague on the behavior of a Hymenoptera, which uses the burrow dug by another to store food needed for its own larvae.[92]

The two men corresponded, and in 1871, in *The Descent of Man*, Darwin recounts Fabre's description of a battle between male Hymenoptera (Cerceris) for the possession of a female, and he qualifies Fabre as an "inimitable observer."[93]

The author of *Souvenirs entomologiques* confesses that he has "deep admiration" for Darwin's "noble character" and his "scholarly candor" and qualifies him as a "deep observer."[94] However, Fabre specifies that "the facts as I observe them distance me from his theories."[95] He seems to move closer to Darwin when he sees the sexual cannibalism of praying mantises as an archaic marker he dates to the Carboniferous period and thus to the primary era.[96] However, he imagines behavioral progression could fit in with the idea of step-by-step creation, which has nothing to do with Darwin's theory of descent with modifications.[97]

Fabre's antievolutionist ideas are marginal, even during his time. Other authors such as the anarchist geographer Pierre

Kropotkine and the previously mentioned Belgian poet Maeterlinck are more favorable to Darwinian theory.

Although Maeterlinck may not be totally convinced of the
truth of the transformist theory, he nevertheless thinks that there
is no better theory and that it is better than having no theory at
all.[98] Kropotkine is more decidedly an evolutionist and admires
Darwin, while pushing his theories further. In *L'Entr'aide, un
facteur de l'évolution* (Mutual aid, a factor of evolution) he dedicates his first two chapters to "Mutual Aid Among Animals"
before detailing the beneficial effects of cooperation among
humans, from medieval times to modern-day workers' associations. Ten or so pages cover social insects.[99]

In the case of bees, the most significant manifestation of evolution is the gradual passage from a solitary lifestyle—still the
rule among the majority of bee species—to a social lifestyle,
demonstrated quite spectacularly in hives. Maeterlinck draws a
parallel between this evolutionary development and that of
humans. Both seem to evolve toward a larger organization, offering individuals more security, but also less independence.[100]

This opposition between individual desire and collective
security, as applied to insect societies as well as to human societies, is integrated into the theoretical framework of psychoanalysis. In 1927 Sigmund Freud receives a letter from a certain
L. R. Delves Broughton. The letter is translated into French the
same year by Marie Bonaparte (1882–1962) for the *Revue Française de Psychanalyse*. A German translation of the letter is done
for the review *Imago* in 1928. It was recently uncovered and analyzed by Remy Amouroux in an article published in the journal
Gesnerus.[101] The letter's author pulls most of his information
from Maeterlinck's *Life of the Bee* and *Life of the Termite*, hypoth-
esizing that hives and termite colonies function with mechanisms of sublimation, allowing workers to accept an asexual life

in order to be useful for the group. He pursues his use of psycho-analytic vocabulary to support this hypothesis (libido, primitive horde, anal stage, regression, identification, etc.).

In a very different conceptual universe, Henri Bergson high-lights the contrast between the evolution of insects and that of men. He broaches the subject as early as 1907 in *L'Évolution créa-trice* (Creative evolution). Bergson is best known for being a philosopher of introspection and intuition. His work reveals a certain admiration for mystics, and one can imagine him suspi-cious of science. In reality, recent historical research demonstrates that not only is Bergson careful to make room for scientific debate in culture, his philosophy is also attentive to the sciences, and particularly to the life sciences.[102]

L'Évolution créatice's pages dedicated to insects reveals he has knowledge of naturalists through first-hand reading of their work. They contain several references to *Souvenirs entomologiques*, with regard to the digger wasps that paralyze insects to use them to lay their eggs.[103] Bergson stands apart from Fabre more clearly than Maeterlinck, although also tacitly. First of all, in this spe-cific case of digger wasps, he indicates that other authors have observed the wasps sometimes killing the caterpillars or half-paralyzing them.[104] Then, and above all, he considers transform-ism to be the right theoretical framework for considering the history of living beings. For him, this history is marked by two major bifurcations, one separating plants from animals, which occurred first, and then one separating arthropods from verte-brates. For Bergson, the first line, which tends toward instinct, culminates with the Hymenoptera; the second line, which tends toward intelligence, culminates with humans.[105] A few years later, in a conference Bergson gives at Birmingham University in 1911, published as *L'Énergie spirituelle* in 1919,[106] he addresses the relationship between the individual and society:

Society, which is the community of individual energies, benefits from the efforts of all its members and renders effort easier to all. It can only subsist by subordinating the individual, it can only progress by leaving the individual free: contradictory requirements, which have to be reconciled. With insects, the first condition alone is fulfilled. The societies of ants and bees are admirably disciplined and united, but fixed in an invariable routine.[107]

Bergson is particularly attached to this theme, and he again opposes insect and human societies in his 1932 *Les Deux sources de la morale et de la religion* (The two sources of morality and religion): "We must not overdo the analogy; we should note, however, that the hymenopterous communities are at the end of one of the two principal lines of animal evolution, just as human societies are at the end of the other, and that they are in this sense counterparts of one another. True, the first are stereotyped, whereas the others vary; the former obey instinct, the latter intelligence."[108]

As a result, we could put forth that Bergson avoids the myth of linear progress, as he gives social insects their place.

Despite the popularity of Bergsonism, the idea that the evolution of the animal world was not awaiting humanization but was also paving the way for Hymenoptera does not gain any ground. On the other hand, the study of animal social behavior, known as sociobiology, raises other debates and controversies.[109] Opponents of sociobiology dispute the use of the word *society* when speaking about animals. The Belgian philosopher Jacques Ruelland says that a society could only exist if "it was conceived in the mind of its members,"[110] and therefore he refuses to refer to a society when describing a simple group of animals of the same species that appeared to be organized.

ANIMAL SOCIETIES?

These controversies are part of the tumultuous history of research and positions over social insects. This history reminds us that there is nothing evident about the matter. How can one consider the integration of humans into the living world without sacrificing their essence? In other words, how can one avoid anthropocentrism without falling into anthropomorphism? Wouldn't using the word *society* to describe insects open the door to recurrent anthropomorphism? Would doing so when commenting old entomological texts be giving into anachronism? To this final criticism, one can note that the use of the word *society*, at least in this context, has been established for a long time. It is already found in Réaumur's work.[111] Latreille also uses it in the first words of his *Essai sur l'histoire des fourmis de la France*. "Of all the insects, the most interested and the most worthy of our research are those that live in society. This kind of civilization requires broader faculties and a particular industry that cannot be found among nomadic insects."[112]

The use of the word *civilization* with regard to insects is a metaphor, as underlined in the original French—*"cette espèce de civilisation."* "Nomadic insects" are those that do not belong to a city. This remains metaphoric, calling largely upon the imagination. It comes as no surprise, then, when a writer during the French Revolution who goes by the name of Michel Dorat-Cubières writes an essay titled "Les Abeilles, ou l'heureux gouvernement" (Bees, or the happy state), in which he says, "To us, bees are like clouds: everyone sees what he wants to see in them."[113]

On the other hand, a century later, *Vocabulaire technique et critique de la philosophie* (Technical and critical vocabulary of philosophy) by the French philosopher André Lalande, provides conceptual clarification by defining society as "a group of individuals among whom there exist organized relationships and reciprocal services." In the same article, Lalande refers to the same distinctions used by Espinas to distinguish different kinds of animal societies: societies of nutrition, reproduction, and relations.[114] Using Espinas's thought as a basis, Lalande falls into a functional more than an anthropomorphic perspective on social phenomena as a biological fact.

During the same time period, the sociologist Émile Durkheim writes: "Animals either live outside of any social state or form simple societies that function thanks to mechanisms that each individual carries inside fully formed since birth."[115]

Recent studies on the social behavior of mammals and birds call into question the supposed simplicity of animal societies. However, one can consider that an interest has been established in first defining what human societies and animal societies have in common and then following up by determining what differentiates them. The disadvantage of this approach are the far-flung metaphors. August Forel himself, whose monumental description of the *Monde social des fourmis* (The social world of ants) is published in 1923, combines, in a rather worrisome manner, psychiatry, entomology, eugenics, and social reform.[116] In 1926 Eugène-Louis Bouvier, a professor at the Muséum national d'histoire naturelle, publishes an essay entitled "Communism Among Insects." He points out that "communist insects" live "without leaders, without guides, without police, without laws in admirably coordinated anarchy."[117] In 1945 the author Geo Favarel, a former colony administrator, publishes *Démocraties et dictatures chez les insectes* (Democracy and dictatorship among insects), in which he opposes termite dictatorship and the "outrageous and greedy nationalism" of wild hives to the "loyal, healthy and confident organization" of our "ant republics."[118]

We find the desire to preserve the specificity of human society, while inscribing it in the larger set of animal societies, in Émile Benveniste's writing. His question is whether the notion of language should be reserved for the human species, to the exclusion of all other animal species.[119] The communication system used by bees serves as a possible counterexample. Summarizing the research of Karl von Frisch in this field, Benveniste evokes the behavior of the bee that returns to the hive "after discovering loot." He describes the dance by which she informs her companions of the resource's location. Further relying on the work of the Austrian entomologist, Benveniste distinguishes a circular dance, corresponding to a close resource, and a figure-eight dance, corresponding to a distant resource. He explains how the rhythm of these movements is a function of the distance and how the axis of the eight indicates the direction to take in relation to the sun. Benveniste concludes: "Bees appear capable of producing and understanding a real message that includes several bits of data. . . . What is remarkable is, first of all, that they demonstrate a capacity to use symbols: there is a 'conventional' correspondence between their behavior and the data it translates."[120]

However, Benveniste is quick to specify that "all of these observations bring to light the essential difference between the communication procedures discovered among bees and our language." Benveniste proposes that bees only use a "code of signals," and he underlines that this code, unlike human language, cannot be broken down into elements that themselves can be broken down.

Although in the end the bee dance does not possess all the characteristics of language, in the eyes of the linguist, this means of communication is characteristic of an insect species "living in society."

5

INDIVIDUAL INSTINCT AND COLLECTIVE INTELLIGENCE

I n *Capital*, Karl Marx defines human labor by contrasting it with the activities of spiders and bees. The former "conducts operations that resemble those of a weaver," and the latter "puts to shame many an architect in the construction of her cells." However, for Marx, human activity remains superior by being thought out before being carried out:

> But what distinguishes the worst architect from the best of bees is this, that the architect raises his structure in imagination before he erects it in reality. At the end of every labour-process, we get a result that already existed in the imagination of the labourer at its commencement. He not only effects a change of form in the material on which he works, but he also realises a purpose of his own that gives the law to his modus operandi, and to which he must subordinate his will.[1]

Associated here by Marx, spiders and bees occupy very different places in the classification of arthropods, as we have seen. Spiders are an order of the class of arachnids, and they are not considered insects. Bees, whether they are solitary or social, form a family, included in the order Hymenoptera, an order itself included in the class of insects. But even more than by their taxonomic status, spiders and bees differ in their symbolic value. Marx's formula evokes Francis Bacon's aphorism 95, in the *Novum Organum*, published in 1620. The English philosopher uses bee activity as an image of the labor of philosophy that "extracts matter from the flowers of the garden and the field" and then "works and fashions it by its own efforts." This action of transformation contrasts with that of spiders who "spin out their own webs," just as it contrasts with the attitude of "empiricists" who, like ants, are content to "heap up and use their store."

THE SPIDER AND ITS WEB

Not all spider species spin webs, and, among those that do, not all produce the exquisite regularity we are familiar with. Studies by the pharmacologist Peter Witt in the 1960s demonstrate that this regularity can be disrupted by placing spiders in contact with caffeine, alcohol, and other drugs.[2] Yet from these fragile constructions spun from matter extracted from the spider's own body, fantasies and stories flourish.

The most celebrated spider-related legend in Greco-Latin tradition is undoubtedly the story Ovid recounts in book 6 of *Metamorphoses*. Arachne, a simple mortal, prides herself on being a better weaver than Minerva herself, and as a result falls prey to the goddess's anger, being transformed into a spider.[3]

Friedrich Christian Lesser, author of *Insecto-Theology*, provides an entirely different perspective. It is not surprising that this work, translated into French in 1742 and into English in 1799, includes spiders, which were then considered to be insects. The author insists on the geometric regularity of spider webs as proof of the existence of a great Geometrician.

A century later, Fabre also attempts to make this geometric regularity an argument in favor of the existence of a God creator. Several chapters in the ninth series of *Souvenirs entomologiques* are dedicated to garden spiders, a kind of spider well represented among European fauna. Chapter 10 discusses web building. Fabre affirms that spider's work is a "polygonal line" woven in a "logarithmic spiral."[4] He is rather quick in his interpretation, as he is to make a connection with the spirals of the ammonite and the nautilus. A logarithmic spiral drawn on a plane has a spacing between each turn that increases geometrically. On the other hand, in a spider web the outer turns seem equidistant and only the central part resembles a logarithmic spiral. The history of geometry attributes the definition of a spiral with equidistant turns to Archimedes and that of a logarithmic spiral to Jacques Bernoulli.[5]

Yet for Fabre the goal is to give readers a feeling of the complex formulas a mathematician would need to describe a structure the spider creates spontaneously. The author's metaphysical intentions become clear in the final lines of the chapter: "This universal geometry tells us of a universal Geometrician."[6]

Whatever the theological backdrop, Fabre's taste for mathematics is no less real. He loves to find calculations and figures in the shapes of nature.[7] In this regard, one could be tempted to compare some parts of *Souvenirs entomologiques* to the work of D'Arcy Thompson. The latter concludes *On Growth and Form* with praise of "that old man eloquent" whom he sees as a "very great Naturalist indeed," who "conjoined the wisdom of antiquity with the learning of to-day" and who "being of the same blood and marrow with Plato and Pythagoras, saw in Number 'la clef de la voûte.'"[8] In reality, D'Arcy Thompson generously lends his philosophical conception to Fabre, and nothing better shows the gap between the two than the diametrically opposite positions they adopt concerning the shape of bee cells.

BEES AND THEIR HONEYCOMBS

The ability of bees to build honeycombs (or cells) to hold their brood and store honey, and to give them a characteristic geometric shape, has sparked research and debate marking the history of mathematics as much that of biology.[9]

These uniform cells are hexagonal in shape. We know that hexagons, like squares and equilateral triangles, make it possible to produce regular paving that, by definition, leaves no gaps between the identical shapes that constitute it. We also know that with equal perimeters, a hexagon delimits a larger area than a square or an equilateral triangle.

From this follows the idea that the hexagonal shape of honeycombs enables bees, for equal volume, to save on the wax used to make the walls. Pappus of Alexandria, one of the last of the great geometricians of antiquity, underlines this point in the preface to book 5 of his *Mathematical Collection*.[10] Fifteen centuries later, the botanist and theologian John Ray, seeking to highlight divine wisdom, explicitly refers to Pappus's demonstration.[11] Honeycombs are not only remarkable for their geometrical form but also for the constancy of their size. At a time when the scholarly community is feeling the need for a universal system of measurements, Réaumur proposes to adopt bee cells as a standard. He regrets that the ancients had not had this idea, because it was "more than probable" that bees "today do not make cells larger or smaller than those made by bees who worked in the time when the Greeks and Romans were the most famous." And he adds: "As Swammerdam tells us, Mr. Thevenot also thought to take a fixed measurement in bee cells."[12] Melchisédech Thevenot was a seventeenth-century French diplomat and traveler known for his scholarship and interest in science. His position on metrology is finely analyzed in a recent article.[13]

Moreover, building a cell also requires building its bottom. Cells arranged on both sides of a wax ray, and placed head to tail, are separated by shared partitions. Observation reveals that the bottom of each cell is a pyramid formed of three lozenges (also called rhombs), and that each cell shares a partition of this pyramid with three neighboring cells having their opening on the other side. The value of the angles of the rhombuses determines the more or less pointed form of the pyramidal bottom of the alveoli.

Giacomo Filippo Maraldi, an astronomer of Italian origin working at the Observatoire de Paris, does the measurements. In 1712, he presents "Observations on Bees" to the Royal Academy of Sciences, in which he indicates the value of 70 degrees for the two acute angles and consequently of 110 degrees for the two obtuse angles.[14]

These figures became crucial, thanks to a meeting between Réaumur and a young German mathematician Samuel Koenig. The latter writes in the *Journal Helvétique* of April 1740:

Madame du Chatelet, Monsieur de Voltaire, and myself saw Monsieur de Réaumur a few days before in Charenton. This skillful savant showed us the artificial hives, which he uses to oblige bees to reveal the secrets of their republic. What he is going to publish on the economy of these animals is admirable, and should astonish scholars as well as the ignorant. The conversation we had that day, having led us to greatly admire the regularity of the little hexagonal boxes, where the bees put their food, and their young, and which are called alveoli, Monsieur de Réaumur took advantage of the occasion to suggest to me a not too difficult, but quite curious problem, to know if bees build their cells in the most perfect way, the most geometrical, and if, of all the possible figures, they chose the one with the most space in the cell, however, they use as little matter as possible.[15]

The scene seems to be drawn from some historical reconstruction including one of the most famous French writers, a woman of letters as seductive as she is scholarly, visiting the author of the *Mémoires pour servir à l'histoire des insectes* and the narrator who is entrusted with a problem of optimization. In the rest of the article, Koenig, who obviously wants to keep up the appearance of a worldly lightheartedness, does not give details of his calculations. He only indicates that having solved the problem "by the *maximis et minimis* method," he calculates that the obtuse angle of the rhombus is 109 degrees 26 minutes and the acute angle 70 degrees 34 minutes. Stressing the almost perfect concordance between the observed angles and the calculated angles, Koenig marvels that an insect, which obviously knew nothing of new mathematical methods, produces "with astonishing precision the result of these same calculations which are unknown to it."[16] And he does not hesitate to infer that "some superior geometrician, both intelligent and wise, presides over this work."[17] He sees in this example the confirmation of a "remark" made by Leibniz on "the consideration of final causes" that is "insufficient to reveal the mechanics of the effects," but constitutes "a principle of invention, which serves without fail to discover how Nature has acted, whenever there is a shorter, a lesser, a better one."[18]

Fontenelle, as perpetual secretary of the Academy of Sciences, is obliged to report on Koenig's work. He does so with humor, and not without slipping in a thought about the superiority of human reasoning: "But in the end, bees would know too much and an excess of glory would be their ruin, we must return to an infinite intelligence that has them acting blindly under orders, without granting them any light that can grow and strengthen on its own, which are all the honor of our reason."[19]

Fontenelle draws a lesson from this discovery, emphasizing that "the principle of final causes, of the best, of the shortest time, way, etc., can be useful in physics," provided that it serves only to "give rise to happy conjectures," which remain to be verified. He notes that this principle, which shows us bees acting economically, must be used with extreme economy.

Fontenelle thus sketches a critical reflection on what today we would call the heuristic and not demonstrative value of finalism. He thus confirms the philosophical significance of the debates to which bee honeycombs gave rise. Half a century later, François Huber also notes that the subject constitutes a favorable ground for final causes and underlines the double dimension—emotional as much as intellectual—of this question of bee architecture: "The order and the symmetry that reigns in their radii seem to invite, in and of themselves, research that satisfies both the heart and the mind."[20]

However, although Huber welcomes the work done by Koenig, Maraldi, and others, he doubts that their solution is rigorously applicable "to the work of these insects." Above all, he regrets that in assuming such strict economy among bees, one has imparted "somewhat narrow views" to nature.[21] He notes that "modern geometers" no longer play a secondary role in the wax economy, and he rejoices because this conception coincides better with the "liberal views of the author of nature."

While Huber thus distances himself from a representation of a God who would save on little bits of wax, this does not mean he was refusing providentialism. His criticism is far removed from what Buffon developed as early as 1753 in *Discours sur la nature des animaux*. This text, which is part of volume 4 of *l'Histoire naturelle*, is mentioned in the introduction as a charge against Reaumur. Beyond the confrontation of two rival scientists, it is the secularization of science that is at stake.

For Buffon, the social life of bees and their common activity results from nothing other than "the universal mechanism and the laws of motion established by the Creator."[22] God has not yet been completely eliminated from the scientific field, but his role is reduced to guaranteeing the relevance of the mechanistic model. No intention, divine or animal, presides over insect constructions. Imagine, Buffon asks us, ten thousand identical "automatons" moving in a "given and circumscribed space." He proposes to equip each of them with a capacity, however minimal, to "feel its existence."[23] These ten thousand individuals would produce uniform and proportionate work, simply because each one of them would have sought to do things in the most convenient way for that individual, while being forced to do so in the way that was least inconvenient for the others.

This mechanical simulation is in fact only a thought experiment that is totally unachievable at the time. Fearing that he would not be convincing enough, Buffon continues on the same topic:

> Would I dare say another word: these bee cells, these praised hexagons, so admired, provide me with one more proof against enthusiasm and admiration: this shape, as geometric and uniform as it seems to us, and that it is indeed in speculation, is only a mechanical and rather imperfect result which is often found in nature, and which we notice even in the roughest productions, including crystals and several other stones, some salts, etc., which consistently take on this shape in their formation.[24]

Appealing again to the imagination, Buffon invites the reader to imagine a container filled with peas or some other cylindrical seed, which would be filled with water, sealed, and then brought to a boil. We would thus obtain, according to Buffon, "six-sided columns." The explanation is "purely mechanical." The seeds become hexagonal "by reciprocal compression" that they exert on each other. In the same way, each bee seeking to occupy "as much space as possible in a given space" must therefore necessarily shape their alveoli to be hexagons, and this "for the same reasoning of reciprocal obstacles."

Buffon's mechanistic explanation supposes that difficult experiments are carried out with uncertain results. The main objection is that bees are not in the situation of seeds in a hermetically sealed container. They are active, and their behavior is decisive.

Another approach is the one proposed in *On the Origin of Species.*[25] Darwin is not outdone in matters of wonder. He has accents that would almost bring to mind a theology of nature, were they not accompanied by the desire to give such phenomena a scientific explanation. His explanation is based on gradual evolution, ranging from bumblebees storing their honey in their old cocoons and supplementing them with "short tubes of wax," to honeybees and their astonishingly uniform cells. To demonstrate the possibility of this gradation between cylindrical cells and hexagonal alveoli, Darwin calls on the reader's geometric imagination. He asks that we picture a large number of identical spheres set out in two layers, the centers separated by a distance equal to the radius multiplied by two. He affirms that "if the planes of intersection between the several spheres in both layers be formed, there will result a double layer of hexagonal prisms united together by pyramidal bases formed of three rhombs." Darwin adds that the angles would be identical to "the best measurements which have been made of the cells of the hive-bee."[26]

In this perspective, the cylindrical cells of a Mexican bee, *Melipona domestica*, appear as an intermediate stage. He could then imagine what modifications of *Melipona* instincts would allow it to build structures similar to those of the honeybee.

In this whole study of bee cell construction, Darwin draws on his own observations and those of other naturalists. He reports experiments that he has carried out, for example, by providing bees with red colored wax to identify the order in which their constructions are made. He thanks several naturalists for their suggestions, including a specialist in crystallography, William H. Miller, professor in Cambridge, who guides him in his geometric approach and even shows him the pattern needed to cut out a dodecahedron.[27]

Fabre reacts strongly to the Darwinian approach. The author of *Souvenirs entomologiques* gives bees and ants relatively little treatment, obviously preferring insects living alone or in small family groups to those who cohabit in the thousands.[28] Is it a simple subjective preference or an expression of political mistrust vis-à-vis large cities? In either case, it is all the more significant that the few pages Fabre devotes to social insects relate to their geometric constructions; without naming the authors, he sums up Buffon's explanation and Darwin's, insisting on what he considers to be their insufficiency.

Where D'Arcy Thompson reviews the various attempts at explanation, preferring a physical approach to the genesis of forms, Fabre refers to an order of the world, which is in his eyes what represents the emanation of divine intelligence. This drift from scientific theory to theological interpretation could be found in many authors in their conception of instinct.

DIVINE INSPIRATION OR NATURAL SELECTION?

Friedrich Christian Lesser, author of *Insecto-Theology* mentioned previously for his thoughts on spider webs, does not hesitate to compare insect instincts to the talents of Bezalel, the biblical artisan of the Tabernacle.[29] "May we not say, without exaggeration, that God hath acted with regard to insects as he formerly did to Bezaleel."[30] With this analogy, Lesser suggests that merit for these constructions is not to be attributed to insects themselves, but rather to the skill and knowledge they receive from the Creator. This theological interpretation introduces a supernatural element—divine decision—and as a result is no longer scientifically admissible. However, acknowledging this nonadmissibility is not enough to determine what science means by the term *instinct*.[31]

The author of *On the Origin of Species*, while apologizing for not defining instinct, qualifies as instinctive any action that requires experience and learning for us to perform, but that animals perform when they are young and inexperienced. The other criteria to qualify as instinctive is that it be performed in the same way by many individuals, without knowing the objective.[32] To complete this description, Darwin emphasizes the similarities and differences between "instinct" and "habit": "If we suppose any habitual action to become inherited—and I think it can be shown that this does sometimes happen—then the resemblance between what originally is a habit and an instinct becomes so close as not to be distinguished."[33]

He goes on to illustrate the difference between instinct and habit with a fictional example: "If Mozart, instead of playing the pianoforte at three years old with wonderfully little practice, had played a tune with no practice at all, he might truly be said to have done so instinctively."[34] In individuals, instinct presents itself as an unlearned action, while it is also what characterizes a species by giving rise to shared behavior. Darwin hypothesizes that "slight modifications of instinct" could be possible; natural selection could play the same role for instinct that it plays for anatomical structures. That would mean a need for "continually accumulating variations of instinct to any extent that may be profitable." Thus it would be impossible to believe that the extraordinary instincts of bees and ants could have been acquired in the span of one generation. This detail is implicitly directed against Lamarck, who Darwin mentions again, explicitly, at the end of the chapter. He notes that supposing that worker bees passed on skills they acquired through heredity runs up against the obstacle of those worker bees being sterile. This objection is less embarrassing for Darwin, who hypothesizes natural selection favoring hives with queens capable of begetting skilled workers.

In the end, habit is to instinct what use and nonuse is to anatomy. This has the same Lamarckian resonance. It has the same limits to that resonance. The same production of a variety of forms. The same role of natural selection. Habit and usage are subjected to the random variations of natural selection, and the same unknown causes lead to anatomic variations and variations in instinct, upon which selection acts.

Darwin is careful not to provide a response to what natural theology views as divine inspiration and what Fabre calls the "ingenuity of the beast."[35] But what all these approaches bring to light is the specific and individual character of instinct. During the twentieth century, a new expression of it arose: deposited in each member of the species, instinct ensures societies' consistency such that they appear to individuals as multicellular organisms appear to the cells that compose it. Several authors develop this analogy in the 1920s.

INDIVIDUALS AND SUPERORGANISMS

Eugene Marais, a South African poet and naturalist, strives to "prove that, like man, the termite nest is an animal composed of instincts," which only lacks "the ability to move."[36] Although

Marais does not appear in Maeterlinck's bibliography, the latter's *The Life of the Termites* invites readers to consider the hive, the anthill, and the termite mound as living beings.[37] In the same years, William Morton Wheeler, the American entomologist already mentioned for his discovery of a manuscript by Réaumur on ants, uses the expression *superorganism*.[38] In 1926 Wheeler writes about insect societies: "These are certainly discrete entities, like the single Metazoon or Metaphyte in possessing a definite boundary, stature, structure and ontogeny and consisting of polymorphic, mutually dependent elements. They may therefore be called superorganisms and constitute a very interesting intermediate stage between the solitary Metazoon and human society."[39]

The notion of superorganism can be found in Edward O. Wilson.[40] Describing the life of a colony of leafcutter ants (*Atta cephalotes*), he explains that the queen does not command, that the organization of social life is distributed among the brains of the workers, and that these separate programs adapt to each other to form a balanced whole. He states that each ant automatically performs certain tasks and avoids others according to its age and size.[41] Wilson also adds that the brain of the superorganism is the whole society; the workers are roughly analogous to the nerve cells.

The same text containing this analogy holds a competing mechanistic vision that explains the title of the article: "Clockwork Lives of the Amazonian Leafcutter Army."

Because social insects are assimilated to superorganisms, they can be perceived as actors in the processes and not only as a theater of the processes through which they take place. However, despite this gain in intelligibility, the superorganism theory has not kept all its promises. In a 1974 review article published in the journal *Communication*, Rémy Chauvin wrote that, "After having gone through a lot of trouble to show that it was, at least, likely, I must admit that it is not as satisfactory in all cases."[42]

The more complex notion of stigmergy, created in reference to termite behavior, also proves to be more operational. Indeed, regulating construction activities represents a real problem for termites, as for all social insects. The theory of stigmergy, from the Greek *ergon*, work, and *stigma*, mark, offers an answer that combines the individual character of the insects with the appearance of coordination.

It involves indirect communication that regulates termite construction activity and involves pheromones,[43] chemicals that allow for communication between individuals a bit like hormones allow for communication within an organism. In a termite mound under construction or repair, a termite places a pellet of earth loaded with pheromones that incites a second individual to bring a second pellet and to pose it on the first rather than anywhere else. Thus, thanks to the accumulation of earth pellets, a pillar gradually rises. The French naturalist Pierre-Paul Grassé proposes the concept of stigmergy in an article published in 1959 entitled "La reconstruction du nid et les coordinations interindividuelles chez Bellicositermes natalensi et Cubitermes sp.: la théorie de la stigmergie. Essai d'interprétation des Termites constructeurs" (Nest reconstruction and interindividual coordination in *Bellicositermes natalensi* and *Cubitermes sp.*: The theory of stigmergy. Essay on interpreting termites builders). Grassé writes: "The coordination of tasks and the regulation of buildings do not depend directly on the workers, but on the constructions themselves. *The worker does not direct its work, but is guided by it.*"[44]

The theory of stigmergy's scope has not been limited to termites. For several authors, this is a fundamental process that illuminates the paradox of social insects: how do animals with such elementary individual behaviors manifest such elaborate collective behavior? The question not only applies to buildings but also to choosing a route. An experiment done by Jean-Louis Deneubourg and his team in Brussels has become famous. Eric Bonabeau and Guy Theraulaz summarize it in *Pour la Science:*

> Argentine ants, *Linepithema humile*, were separated from a food source, accessible via two routes, one twice as long as the other. In a few minutes, they choose the shortest. How? The ants follow routes marked by a pheromone, which they leave behind. The first ants to return to the nest from the food source used the shortest route in both directions, and the path was thus marked with the pheromone twice, attracting more of the other ants than the longer path marked only once.[45]

Similarly, Deborah M. Gordon's work focuses on the processes that determine the number of workers assigned to different tasks (harvesting, larval care, construction and maintenance of the anthill).[46] These staff are interdependent and evolve from day to day, even hourly, in response to needs. An individual's choice to engage in a particular task is determined solely by age or genotype but is also influenced by the encounters it makes. As in the case of termites building or repairing their termite mound, it is a kind of collective intelligence or, as some authors call it, swarm intelligence.[47]

The way in which intelligent collective behavior emerges from elementary individual reactions has given rise to the idea of analog modeling in which ants are replaced by small robots.[48] The situation evokes Buffon and his ten thousand "automatons" who are supposed to produce work as regular and proportionate as that of bees. However, and the difference is great, what was for Buffon a thought experiment has now become a carried-out experiment.

The construction and use of robots initiates a process of denaturalization, the next step being the appearance of a virtual ant, made possible by advances in computer science.

This allows the processing of a problem that dates from the middle of the nineteenth century, which we find formulated along with a brief history in an article by the American mathematician George Danzig released in 1954. It is a classic optimization problem: the itinerary taken by a traveling salesperson who must pass through different cities, once and only once. Marco Dorigo and his colleagues approached the problem using a computer simulation of ant behavior by randomly releasing "independent virtual ants" on a network of cities, each depositing a certain amount of virtual pheromone on the routes they took. The operation is repeated several times and, as stated in the aforementioned *Pour la Science* article, as the tests progressed, "the paths taken by the artificial ants were reduced and, ultimately, the connections they favored, when put end to end, constituted a short overall path."[49]

Artificial intelligence specialists use what they call "ant colony algorithms" of this type to solve more than just the problem of the traveling salesperson. Social insects are thus summoned to solve problems of routing, distribution, and logistics. The result is scholarly studies with unexpected title such as "Ant Colony Optimization for the Problem of the Multidimensional Backpack."[50] This mobilization remains in the domain of simulation, and some entomologists might consider it a contribution of entomology to computer science rather than a proper development of entomology. In addition, it should be noted that the problems in question are essentially the search for an optimum. This is the case at the time when Koenig and Fontenelle plead for a circumspect use of final causes in the study of bees' cells. *Mutatis mutandis*, it is still so today. The lesson we could draw from this line of research is to recognize the heuristic role of an analogy without drawing hastily a metaphysical or moral conclusion.

Deborah M. Gordon also ends an article published in *Nature* called "Control Without Hierarchy" with a warning: "Life in all its forms is messy, surprising, and complicated. Rather than look for perfect efficiency, or for another example of the same process observed elsewhere, we should ask how each system manages to work well enough, most of the time, that embryos become recognizable organisms, brains learn and remember and ants cover the planet."[51]

The desire to determine under what conditions analogies between societies of insects and other self-organized entities could be fruitful can be found in a recent work by Henri Atlan, *Le Vivant post-génomique ou Qu'est-ce que l'auto-organisation?* (Postgenomic life, or What is self-organization?]. Atlan, a biologist attentive to the philosophical scope of science, known for his work modeling self-organizational processes, evokes the "'intelligent' collective behavior" of social insects, as seen in the construction "of sometimes very architecturally complex hives and anthills." He recalls "that swarm intelligence is a collective intelligence that emerges from a group of simple agents" and he adds: "This is why . . . the transposition of these collective behaviors to human societies can not be made without critically analyzing the assumptions about the individual behaviors of the agents."[52]

As Henri Atlan himself points out, this type of analysis does not apply to all human behavior, because much individual behavior "is neither simple nor trivial," but these models are relevant to study "collective human phenomena such as, for example, traffic or crowd movements."[53]

Thus these models have circulated, models that one might be tempted to suspect of anthropomorphism, yet can also appear as humans making conceptual use of insects.

6

BATTLES AND ALLIANCES

I f mosquitoes and other vile parasites were skilled at the art
of discussion," they would perhaps decide "that man was
created to feed them with blood." This mindful thought
springs from the imagination of Émile Blanchard and resembles the start of a philosophical fable, capable of overturning the
false simplicity surrounding the ideas of useful and harmful
insects. However, Blanchard quickly abandons this audacious
story, returning to an approach that conforms more readily to
the expectations readers had of a work of entomology in the nineteenth century. He considers that man had the right to defend
himself and thus it is "necessary" that he seek to destroy the
"creatures that attack him" and rightful that he seek to protect
his harvests.[1] He insists that, in this battle, man has available to
him a simple approach that consists of "knowing his enemies."
In other words, entomology provides strategic knowledge. As for
useful species, he defined them as those that "kill harmful species" or that could provide us with "products" such as dyes, drugs,
and, of course, honey and silk. A century after Blanchard, the
social utility of entomological expertise has proven to be occasionally unexpected. Thus the succession of insect species that
settle in a corpse allow forensic entomologists to determine a
time of death.[2]

HONEY, WAX, AND SILK

"What is bitter in the mouth is good for the body." With this formulation, Gaston Bachelard summarizes the prescientific representation that tends to link the effectiveness of treatment to the unpleasantness it causes.[3] In view of this common opinion, honey offers a real paradox as both medicine and food—almost candy.

We could also refer to wax as a paradox of reality, another product of the hive, that can take all the forms. To explain and justify the way in which he defines matter by extension, Descartes takes the example of a piece of wax, "having been but recently taken from the beehive," that "has not yet lost the sweetness of the honey it contained" and "still retains somewhat of the odor of the flowers from which it was gathered."[4] Once placed near a fire, the piece of wax begins to melt and its visual, olfactory, and taste qualities change, leading Descartes to conclude: "There certainly remains nothing, except something extended, flexible, and movable."[5] Matter being defined by extension is one of the fundamental theses of Cartesian philosophy and would apply to any material reality. Nevertheless, the choice of wax as an example is important. Changing the shape of a piece of wood or a piece of metal would require a series of gestures, perhaps even the use of a tool. What was singular in wax is that, because of its plasticity, it offers a concrete example of the general properties of matter.

While harvesting honey from wild bees and beekeeping, in its traditional and modern forms, have made honey and wax products sought after and used in a very large number of cultures,[6] on the other hand, the principle of silk production, discovered by Chinese artisans several centuries before the common era, long remained secret. At the end of the Roman era, Europe took a liking to this fabric, which was both luxurious and practical, but did not know how to produce it and had to make at least indirect contact with the Far East in order to obtain it: hence the mythical Silk Road.[7] Instead of using an image of a road, we should use that of a network whose nodes recall images and stories. From the fortifications of Antioch to the Great Wall of China, from the port of Trebizond to the Samarkand mosques, the game of political alliances, the dissemination of religious ideas, and the exploration of unfamiliar lands all followed the paths opened by the silk trade. The star role in these exchanges that span the Eurasian continent is the *Bombyx mori* caterpillar. During the Renaissance, cultivation of the white mulberry tree, whose leaves provide food for silkworms, spreads to Italy and southern France. As Olivier de Serres writes, "Wherever the vine grows, there may come silk."[8] A chapter of his *Théâtre d' agriculture et ménage des champs* (Theater of agriculture and field maintenance) is devoted to the "collecting silk by the food of the worms that made it."[9] Maximilien de Béthune, duke of Sully, a statesman who served King Henry IV of France, promotes mulberry cultivation and silk farming, contributing to creating an image in the mind of the French people of the good King Henry and carrying out one of the first examples of a policy of economic development based on scientific expertise.[10]

Two and a half centuries after Olivier de Serres, the appearance of a disease impacting silkworms causes another intervention by public authorities. Louis Pasteur, preceded by his fame, is responsible for saving French sericulture. He meets Jean-Henri Fabre. He notes that the Parisian chemist still knows nothing about the biology of the *Bombyx mori*, to the extent that, never having seen a cocoon, he waves it in front of his ear and, astonished, says, "Is there something inside? —But yes. —And what is that? —The chrysalis. —What is the chrysalis?" And yet the

Provençal entomologist stresses that Pasteur's observations under the microscope would revolutionize the hygiene of silkworm farms.[11] For Fabre, this anecdote illustrates the preeminence of knowledge acquired through observation—including under a microscope—over prior book knowledge, and yet, beyond this epistemological lesson, it is significant that the call for expertise relates to silk, a product with high economic value.

Finally, it is in this precious fiber that the Lyonnais canuts or silk workers—immortalized by Aristide Bruant—promise to weave "the shroud of the old world." There is the matter for a cultural and social history of the *Bombyx mori*, modeled on Hélène Perrin's articles about the cotton weevil, Charlotte Sleigh's work on the ants, and Yves Cambefort's work on beetles.[12]

In addition, traditional pharmacopoeia includes a number of drugs derived from insects. Recent research has renewed and expanded this entomological contribution to the therapeutic arsenal.[13] Less known, the use of fly larvae (*Lucilia sericata*) to clean wounds has been attested in medicine and veterinary medicine. Neglected after the invention of antibiotics, it has been rediscovered and administered as a dressing for its bactericidal action.

PESTS AND VECTORS

Honey and silk are synonymous with sweetness and luxury, but many insects have been associated with images of havoc and destruction since as far back as ancient times, as evidenced by the biblical account of the Exodus. Among the ten plagues that Yahweh inflicts on Egypt to wrest the freedom of the Hebrew people from the Pharaoh, there are three that involve insects: the third was gnats, the fourth flies, and the eighth locusts. The last is presented as the provoked and paroxysmal manifestation of a usually less catastrophic episode:

> And the Lord said to Moses, "Stretch out your hand over Egypt so that locusts swarm over the land and devour everything grow-ing in the fields, everything left by the hail." So Moses stretched out his staff over Egypt, and the Lord made an east wind blow across the land all that day and all that night. By morning the wind had brought the locusts; they invaded all Egypt and settled down in every area of the country in great numbers. Never before had there been such a plague of locusts, nor will there ever be again. They covered all the ground until it was black. They devoured all that was left after the hail—everything growing in the fields and the fruit on the trees. Nothing green remained on tree or plant in all the land of Egypt.[14]

Inheriting, if only indirectly, biblical criticism, historical science ceases to see such narratives as echoes of real events, but leaves open the possibility that the stories recount how such phenom-ena, which are frequent in this region of the world, were perceived.[15]

Insects can ravage living plants, stored seeds, building mate-rials, textile fibers, and so much more. Most often, and this is illustrated by the biblical story, this competition modifies the relationship of power between human groups. The invasion of European potato crops by the Colorado beetle, a Coleopteran from the American West, marked European memory. Phyllox-era, a *Hemiptera*, also American, durably modified the map of French vineyards.[16]

The network of interactions becomes more complex when the insect transmits a parasite.[17] This is the case of elephantiasis studied by Patrick Manson, a Scottish doctor who practiced in China. This is the case of malaria, which impacts many places.[18] When the French military doctor Alphonse Laveran discovers that the pathogen of malaria is a parasitic microorganism of the genus *Plasmodium*, he opens the way to the recognition of the complete cycle, including the mosquito. Among the writers who have worked in this field, we usually remember the names of the Italian Giovanni Battista Grassi and the Englishman Ronald Ross. The quarrel over priorities that opposes them reveals the difference in their scientific style. Grassi, a zoologist, introduces a concern for naturalistic precision in the determination of the mosquito species concerned. Ross stands out for his experimental and mathematical approach. Calling on the British government to step up the fight against malaria, in 1902 Ross publishes *Mosquito Brigades and How to Organize Them*.[19] One of the chapters is entitled "History of the War Against Mosquitoes," a formulation reminiscent of Bruno Latour: "In war, there were always two enemies, the macroscopic and the microscopic."[20] We can note, however, that malaria must not only face microbes and humans but also insects. The Indian novelist Amitav Ghosh drew a dramatic plot twist from this situation for his historical novel *The Calcutta Chromosome*.[21]

From our point of view, the most profound change is that pest insects are competing for a resource, while the vector insects treat us humans as a resource. One could even push the paradox and consider the human who, contaminated by a first sting, transmits the microbe to the insect as the equivalent of a vector.[22] Indeed, the mosquito-*Plasmodium*-warm-blooded-vertebrate life cycle assumes the insect carrier of the disease must bite first before a second sting allows the insect to infect a second person. It is precisely on the basis of this necessity that Ronald Ross begins to build a mathematical model of the transmission of malaria. In the work already quoted, he imagines that if one halves the number of mosquitoes in a locality that mosquito bites would be halved as well. But, he adds, as the mosquitoes themselves become infected by biting people already infected, the percentage of mosquitoes infected among the remaining insects will be reduced in turn and will be only a quarter of what they were before.[23] This argument, intended for a wide audience, can only give a rough idea of the models published by Ross in the following years that would attract the attention of the American mathematician Alfred Lotka.[24] He is also known in ecology as having discovered, at the same time as the Italian Vito Volterra, and independently of him, the equations that model the fluctuations of two populations, one of prey and the other of predators. In 1935 Volterra and his son-in-law, the biologist Umberto d'Ancona, publish a book entitled *Les Associations biologiques au point de vue mathématique,* in which they note: "The quantitative study of interspecific relationships was first discussed by Ronald Ross with regards to the relationship between humans and malaria-carrying mosquitos. Subsequently, various authors would pursue such research. Ross establishes equations to infer the curve of a malaria epidemic among human populations in relation to the number of bites by infected mosquitoes."[25]

The modeling Ross proposes is a milestone in the history of ecological theory and its relation to applied entomology.[26]

THE ENEMIES OF OUR ENEMY

French entomologist Jacques d'Aguilar recalls that there are three different approaches to fighting insects that are considered to be harmful: physical, chemical, and biological.[27] A physical approach primarily uses temperature and is mainly used for the vegetables in storage. Chemical control involves synthetic products toxic to insects that are injected into the soil or dispersed into the atmosphere. Biological control relies heavily on the use of predators and parasites that attack the insect whose populations one wants to control.[28]

On the subject of chemical control, historian of science Sarah Jansen emphasizes how Germany applied gas warfare technology (used during World War I) to the fight against forest insects. She points out how certain German entomologists of the time came closer to political power through the use of terms such as *purity, degeneration*, and *war* and integrated *military techniques*.[29] In this transfer of the war industry to applied entomology, Fritz Haber, one of the inventors of gas warfare who mobilized the chemical industry, plays a key role. He is so involved that in 1915, when his wife Clara, also a chemist, discovers the extent of his role she chooses to commit suicide, as she disapproves of chemical weapons, seeing them as a barbarity. The controversies over gas warfare do not prevent Haber from being awarded the Nobel Prize in chemistry in 1918 for his work on the synthesis of ammonia, which benefited the manufacturing of both fertilizers and explosives. In addition, Haber's name is also linked to that of Zyklon B, an insecticide, which was used, hardly transformed, by the Nazis in the gas chambers at the extermination camps. Haber was not present for this final phase of the genocide: as he was from a Jewish family, he had left Germany shortly after Hitler came to power and fled to England and Switzerland. He died in Basel in January 1934.[30]

The use of DDT, or dichlorodiphenyltrichloroethane, also begins in the context of the Second World War. It was in 1939 that the Swiss chemist Paul Hermann Müller (Nobel Prize in medicine in 1948) brings to light the insecticidal properties of this compound, which had been known since 1874.[31] Use of DDT made a massive contribution to fighting food shortages that affected the populations in belligerent countries by decreasing the number of insect pests. Similarly, in malaria-affected areas, DDT is part of a fight against the mosquito that vectored disease. In Naples, in December 1943, an epidemic of typhus breaks out. As this disease is transmitted by lice, U.S. forces begin to treat more than two million people with DDT and in March 1944 succeed in stopping the outbreak.[32]

However, the accumulation of insecticides—especially DDT—in food chains results in the concentration of toxins in birds that prey on insects. In 1962 Rachel Carson, a specialist in marine biology and ecology, publishes *Silent Spring* where she imagines the disappearance of insectivorous birds. In addition, the repeated use of an insecticide results in the selection of forms resistant to this insecticide. This selection, which virtually confirms the validity of the theory of natural selection, aggravates the phytosanitary situation.

Today biological control enjoys a better image than chemical control in many circles. However, although it has been successful, it has also experienced failures. Among the failures, the introduction of invasive species has weakened island environments. Among the successes, the work of Charles Riley occupies a prominent place.[33] Born in 1843, of British origin, he is educated in Dieppe, then studies in Bonn, before emigrating to the United States. He works on a farm before being a reporter for a farm magazine in Chicago. He is an entomologist for the state of Missouri and teaches courses in several universities. Both a Francophone and a Francophile, he makes seven trips to France. His contribution to biological control relates to the protection of citrus crops in California. These were ravaged by *Icerya purchasi*, a scale insect accidentally introduced from Australia in 1868. Riley is convinced that if this scale had not proliferated in its country of origin, it is because its populations were limited by predation. Riley manages to send a mission to Australia. This results in the voluntary introduction in America of an Australian ladybug, *Rodolia (Novius) cardinalis*, that reduces *Icerya purchasi* numbers to a tolerable level in less than two years.[34]

In France, Riley's name brings to mind his friendly collaboration with Jules Émile Planchon and other French naturalists in the fight against phylloxera. The idea is to graft French grapes onto American vines that are resistant to the aphids. This episode is the subject of a lively and erudite historical study published in 2007 by the *Annales de la société entomologique de France*.[35] This is another form of biological control; it does not involve a particular predator, but is rather based on the knowledge of the insect's life cycle and its needs, which are used to develop a strategy that is favorable to the plant.

However, the fight against phylloxera remains conceived in the mode of antagonism. The concept of the ecosystem has met with scientific and media success because it highlights nonadversarial relationships, in other words, games where multiple partners can win together. Scavengers, and insects among them, play a decisive role in this approach. The history of dung beetles, which Australian farmers introduce into their countryside to avoid being buried in manure from livestock imported from Europe, has emblematic value. Another significant example is the use of weevils against the invasive aquatic water hyacinth.[36] Even more than recycling of organic matter, pollination results from a tacit agreement between an insect and a plant.[37]

POLLINATION: A SECRET OF NATURE?

The Greeks, who probably inherited such techniques from the Babylonians, already knew that when cultivating date palms, to ensure a good harvest of dates, one needs to shake male flowers over female, fruit-bearing inflorescences.

This could lead one to believe that although the ancients may not have understood the mechanism, they at least admit the principle of sexual reproduction of plants. In fact, to know that there are two forms within a species, one of which is obviously fertile, but the other of which is also required for the production of an offspring, does not automatically lead to a distinction between the two sexes. In this case, a consultation with Theophrastus's *Enquiry Into Plants* reveals that the Greek naturalist knows about artificial fertilization of the date palms, but we are unable to say that he knows about the sexual reproduction of the plants.[38] To grant sexuality to plants is to conceive of sexual reproduction not as a specific case but as a phenomenon that concerns all plants. Such knowledge does not assert itself until the end of the seventeenth century. We owe the first experimental verification of the sexuality of flowers to Rudolf Jacob Camerer, known as Camerarius, a professor of medicine and director of the Botanical Garden of Tübingen, who in 1694 reports on it in a letter with the explicit title of *De sexu plantarum epistola* (Letter on the sex of plants).[39] At the Jardin des Plantes, in Paris, Sébastien Vaillant enthusiastically supports this idea. Linnaeus does the same in his 1729 thesis entitled *Praeludia sponsaliarum plantarum* (Preludes to plant weddings).[40] There, the Swedish botanist explains the function of the different parts of the flower. Plant sexuality seems to him so spectacular that he does not hesitate to use it as a basis for his classification of plants.[41]

Once the need for fertilization has been admitted, it remains to be determined how the approximation of the sexes can take place or, in other words, how the male element, the pollen, contained in the stamens, can be brought into contact with the female element, the pistil. This contact is essential in plants with separate sexes, that is to say, in species that have flowers with stamens and others with pistils. It is also common in species whose flowers have both stamens and pistils, but which, by having their pollen carried on the pistil of another plant, avoid self-fertilization.

The first experiments focus on species with separate sexes and pollen transported by the wind. This was the case of the *Mercurialis annua* used by Camerarius. Observations of insect transport come later.

Arthur Dobbs, governor of North Carolina, is one of the first, when he was living in Ireland. There, future among other "rural amusements," he devotes himself to observing bees. From these observations he draws the substance of a letter to the *Philosophical Transactions of the Royal Society of London* (1750). He explains that he follows in the footsteps of Réaumur's *Mémoires pour servir à l'histoire des insectes*, but that, contrary to what was written in the latter, he uncovers that when bees forage, they do not go from one species of flower to another during the same flight, but rather limit themselves to the flowers of the same species. In his opinion, bees do this because Providence has given them the responsibility to contribute to the plant's development, while acting for the good of their own species, and if bees acted otherwise, they would mix pollens, to the detriment of the plant.[42] Here he implies the principle of pollination and what will be called later the sterility of hybrids.

Philip Miller, who runs the Chelsea medicinal gardens, reports another observation.[43] He carries out an experiment in fertilization using tulips, which he then recounts a few years later:

> I set twelve Tulips by themselves, about six or seven yards from any other, and as soon as they blew, I took out the stamina with

their summits so very carefully, that I scattered none of the males dust; and about two days afterwards I saw bees working on a bed of Tulips, where I did not take out the stamina and when they came out, they were loaded with the farina or male dust on their bodies and legs; and I saw them fly into the Tulips, where I had taken out the stamina, and when they came out, I found they had left behind them sufficient to impregnate these flowers, for they bore good ripe seeds which afterward drew.[44]

In the matter of plant fertilization, the end of the eighteenth century is marked by the research of Joseph Gottlieb Koelreuter and Christian Konrad Sprengel. The St. Petersburg Academy of Sciences offers a prize for a work that would more fully demonstrate the sexuality of plants, and Koelreuter carries out experiments that Darwin readily refers to when he discusses hybridity in chapter 8 of *On the Origin of Species*. Sprengel also attracts Darwin's admiration. In *Das entdeckte Geheimnis der Natur im Bau und in der Befruchtung der Blumen* (The secret of nature discovered in the form and fertilization of flowers), he describes the fine structure of the flowers in connection with the pollination. In particular, he demonstrated the role of nectaries, which are the glands located in flowers that exude a sweet liquid to attract insects.[45]

The discovery of pollination reveals the previously unknown role of many insects, disrupting the separation between harmful and useful insects. It should be noted that this separation did not exclude indirect utility, which was even called for political as well as theological reasons. Take, for example, the 1752 doctoral dissertation by one of Linnaeus's students by the name of Christopher Gedner.[46] This dissertation, entitled *Quaestio historico naturalis: cui bono?*—which can be rendered by "What is natural history good for?"—was, as was customary at the time in Sweden, inspired, even written by, the teacher himself, who was Linnaeus in this case.[47] The author regrets that, for the common people as for the great of this world, the study of nature is a vain curiosity, only understandable when it is applied to large objects but frankly ridiculous when dealing with insects or mosses.[48] In order to fight this prejudice, Linnaeus reminds us that natural history can introduce exotic species, including plants, of course, but also certain predatory insects that fight against insects that disturb us. Above all, Linnaeus invites readers to believe that all created things are useful to us, either directly or indirectly. So that "what seems to us quite harmful, often is most useful to us." For example, aphids are eaten by larger insects who themselves serve as food for passerines, on whom we feast on and who charm us with their songs.[49] In the end, everything happens as if Linnaeus, in affirming the utility of all creatures, justifies the social function of natural history. In other words, insects are necessarily useful—since for Linnaeus the Creator had done nothing in vain—but of a utility that must be discovered, ensuring the utility of the naturalist.

This indirect utility, this delegated value, this refracted beauty is not the only one that can be found in insects. A golden ground beetle, the sail swallowtail, morpho butterflies that animate the tropical forests of the New World with their metallic blue, and a thousand other species contribute to an aesthetic perception of the world. Not to mention dragonflies, about which Alain Cugno shares his wonder in free and precise language.[50]

The division between beneficial insects and pests has never been obvious to the point of preventing new "useful" insects, the pollinators, from appearing. Two centuries later, the fight against insects considered as harmful causes disastrous disturbances in the functioning of many ecosystems. Agricultural production itself is threatened. The damage caused to beekeeping by certain insecticides raises fears of the worst for the survival of bees and in turn for the fruit crops they pollinate. In northern California, for example, farmers are forced to rent hives and bring them by truck, or their production of fruits and vegetables will be greatly reduced.[51] Thus the concept of ecological service takes form. Ecological services are services provided by the functioning of ecosystems. Pollination is in a good position. The fully assumed anthropocentric perspective places the threatened insects in an economic rationality.

A dark prediction, attributed to Albert Einstein and often quoted, states that if the bee were to disappear, humanity would survive only a few years. This deserves discussion. Certainly, the disappearance of *Apis mellifera* would be an ecological disaster that would affect the pollination of many flowering plants, causing the disappearance of a large number of them. This would not only diminish biodiversity but also the quality of life of most human beings. However, not all flowering plants are pollinated by insects, and those pollinated by insects are not all pollinated by bees. In terms of rhetoric, the phrase is a hyperbole. Its attribution to the father of the theory of relativity is more than improbable. Asked about it by a journalist, the curator of the Albert Einstein Archives at the Hebrew University of Jerusalem, Roni Grosz, cautiously responds that it is difficult to say that a quote is false, but that in any case he never encountered any such a sentence in Albert Einstein's writing.[52] Ultimately, the uncertainty of the author of the formula and the criticism of exaggeration should not lead us to forget the danger represented by the erosion of biodiversity.[53] Policies are being implemented to counter this danger. Some experts consider them completely inadequate given the seriousness of environmental issues; others consider them too expensive economically.

One noteworthy example is a move that stopped the construction of a section of motorway between Le Mans and Tours, France, for several years. The reason for this interruption was the discovery, in 1996, of Coleoptera larvae of *Osmoderma eremita*, known as the hermit beetle, on the planned location of the section connecting the two cities. As this beetle species was protected by a European directive, the National Museum of Natural History requested an impact study. The report, issued in 1999, under the responsibility of Patrick Blandin, advised the creation of *Natura 2000* zones,[54] while stressing that "the impact of the motorway would probably be lower than that of the regrouping of agricultural parcels in the areas crossed."[55] Indeed, "the hermit beetle was widely present in oaks planted in the hedges," which required "the diagnosis of all the hedges of the sector and about 17,000 trees likely to harbor Coleoptera."[56] The museum's expertise led to the conclusion that, in view of the measures taken (modification of the infrastructure, creation of *Natura 2000* zones, land consolidation master plan, entomological monitoring), the effects on the habitat of the hermit beetle "would be insignificant overall."[57] The roadwork could resume; the motorway was opened in December 2005. The social aspect of the affair inspired director Xavier Giannoli with a film entitled *À l'origine*, presented at the Cannes Film Festival in 2009, with François Cluzet, Emmanuelle Devos, and Gérard Depardieu among the actors. The action was transposed to the north of France. The landscape, the fauna, and the flora had only a discreet presence, as if, in any case, it was too difficult to mobilize public opinion for a beetle. This difficulty was reflected in a debate in the National Assembly: MP Yves Deniaud was scandalized at the money spent on this "dear, very dear creature."[58]

Neither object of affection, as is the case of some vertebrates, nor eminently useful auxiliary like pollinators, the hermit beetle certainly participates in the recycling of dead wood, but this is a modest task in the eyes of traffic officials. An argument for the protection of this beetle could be based on it being part of the country's heritage, the memory it carries of a farming landscape shaped by centuries of agricultural activities, as well as the procession of associated species of flora and fauna. What is most precious in a species like this, which, beyond the simple respect of the law, justifies the measures taken for its protection, is the environment that conditions its survival and makes its presence an indicator of a rich ecology.[59]

There are two ways of naming and socially valuing natural realities: the notion of ecological service and the concept of common heritage. One is borrowed from the world of economics, the other from the world of cultural property. Both apply very well to insects, so that, between them and humans, a language other than that of war is possible. These rhetorical displacements may seem derisory in relation to shortages and diseases that affect human populations because of certain species of insects or in relation to the threats of extinction that weigh on other species. In fact, to wonder about the means of living with insects is not to attach as much importance to the life of a mosquito as to that of a human, it is simply to seek the optimal conditions for a life of coexistence that takes into account both the different roles that insects have played in human history—by their direct or indirect, secret or spectacular, beneficial or deadly action—as well as the conceptual innovations they have inspired or favored.

7

MODEL INSECTS

In a short story called "Del rigor en la ciencia" ("On exactitude in science"), Jorge Luis Borges dreamed about a "Map of the Empire whose size was that of the Empire."[1] The invention of this paradoxical object that is itself a substitute for itself is reminiscent, for readers of Jean-Jacques Rousseau, of Emile's natural history closet. A museum "richer than that of kings" because it was "the whole world."[2] And, Rousseau specified, "Everything is in its right place; the Naturalist who is its curator has taken care to arrange it in the fairest order; Daubenton could do no better."[3] In other words, for those who want to know nature, the terrain is more instructive than the collection, even if the latter is ordered by the one who, under the responsibility of Buffon, cared for and curated the king's garden's collection. A similar statement, with the same reference to Daubenton, as an emblematic figure of museum curator, is found a decade later in the writings of Bernardin de Saint-Pierre: "What a show do our collections of animals in our museums present us! Vain is the art of Daubentons to give them an appearance of life." Everything reminds us that "the features of death have struck them." In the end, he concludes, "Our books on nature are only the novel, and our museums the tomb."[4]

Paradoxically, life threatens death. Indeed, what the manager of a private or public collection of insects fears most is a form of life that threatens the exposed remains: the dead insect is a food source for other insects. Dermestes, Anthrenus, Psocoptera, and a few others, threaten the fragile corpses that are in charge of representing the species in which they have been placed. To prevent the process of degradation, or at least slow it, the manager of a collection handles poisons and takes multiple precautions.

The constitution of a collection is based on the mastery of a know-how, acquired in specialized publications and transmitted in an academic or associative framework. Moreover, no matter the purchases and exchanges or the size of donations and bequests, catching insects with the required equipment continues to play a fundamental role, hence making for possible contradictions with the requirements of protecting nature.

When an insect enters a collection, it must pass through a series of gestures, described in great detail by Jacques d'Aguilar:

> Harvested insects are prepared, spreading legs, antennae and wings so as to present a unit that helps in comparison. They are poked with special pins, called entomological, then spread out on a board made of cork or of plastazote to hold the needles used to maintain the appendices until dried. . . . Once dry, the samples are classified in "insect boxes" of standardized dimensions, in which paradichlorobenzene, creosote or thymol is introduced to protect the collection from mold and certain insects, such as Anthreni or Psocidae.[5]

All this concerns insects caught in the adult stage (entomologists use the term *imago*); larvae are kept in glass tubes filled with alcohol or other liquid. The management of a collection of insects thus appears as a daily struggle against the untimely life of the pests that constantly threaten scholarly collections.

Beyond their heritage value, all these specimens are an indispensable tool for studying the diversity of living forms. Their very arrangement materializes the classification of the species. Far from losing importance, the collection of many specimens becomes more and more necessary insofar as a population-based conception of species replaces a typological conception: collections tend to testify to the variability within each species as well as diversity between species. But the usefulness of collections is not limited to the systematic; it extends to understanding some of the mechanisms that explain biodiversity. This is illustrated by the example of mimicry.

MIMICRY

Among the insects that flutter in European gardens in summer, we note that there are some that resemble wasps but have only two wings instead of the four found in wasps. Furthermore, they possess neither dart nor venom. A closer look reveals that they are in fact flies, in the family *Syrphidae*. Entomologists interpret this similarity to wasps as a means of protection, insofar as it is supposed to intimidate possible predators. Although found in temperate countries, this type of resemblance, called mimicry, was first studied in the intertropical zone.

It is Henry Walter Bates, a naturalist and explorer who spent several years in the Amazon, who observes the phenomenon in several species of butterflies. In November 1861 he describes it and offers an explanation to the Linnaean Society of London, which is then published the following year in the society's proceedings. The excessively sober title, "Contributions to an Insect Fauna of the Amazon Valley. Lepidoptera: Heliconidae," and the focus on a single family, could lead one to overlook the magnitude of the theoretical issues found in the presentation. Indeed,

Bates criticizes "closet naturalists" who multiply small species, without seeing that they are in fact only varieties. He attributes this tendency to the fact that they work on specimens without any connection to each other. And he adds that in so doing it is not surprising that they have faith in the absolute distinction between species and their immutability.[6] Bates concludes his argument as follows: "Those who earnestly desire a rational explanation must, I think, arrive at the conclusion that these apparently miraculous, but always beautiful and wonderful, mimetic resemblances, and therefore probably every other kind of adaptation in beings, are brought about by agencies similar to those we have here discussed."[7]

Alfred Russel Wallace, a friend of Bates, and his traveling companion in Brazil, prominently features mimicry from an evolutionary perspective. Wallace's fame comes from independently developing the theory of species transformation at the same time as Darwin. We know that the two men, rather than engaging in a quarrel of priority, prefer mutual recognition of their contributions.[8] In 1889 Wallace publishes a book called *Darwinism*, presenting it as an "Exposition of the Theory of Natural Selection with Some of Its Applications." Among these applications is an explanation of mimicry. Before emphasizing the decisive role of Bates, Wallace defines mimicry as "a form of protective resemblance, in which one species so closely resembles another in external form and colouring as to be mistaken for it, although the two may not be really allied and often belong to distinct families or orders."[9]

Wallace shares Darwin's enthusiasm for Bates's explanation of mimicry. In a letter dated November 20, 1862, the author of *On the Origin of Species* writes to Bates, commenting on the latter's article: "it is one of the most remarkable and admirable papers I have read in my life."[10]

In 1879 the German naturalist Fritz Müller, who is studying Brazilian butterflies in the field, discovered another type of mimicry. In this form, now called Mullerian mimicry, two or more toxic prey species look alike, and this similarity allows predators to learn faster to avoid these prey, which decreases the burden of predation for the two species.

Two forms of mimicry, Batesian and Mullerian, could be found in a book published in 1915 under the title *Mimicry in Butterflies* or, in other words, "mimicry among butterflies."[11] The author, Reginald Punnett, marked the history of genetics: he is one of the people, along with William Bateson, to introduce the Mendelian approach in Britain. In the preface to his book he indicates that he is writing for several kinds of readers looking for an illustrated presentation that was not too long, not too expensive, and not too difficult. In this way, Punnett hopes to encourage those traveling or staying in the tropics to observe the numerous and spectacular cases of mimicry found in these countries, thus helping to shed light on what he considered to be one of the most fascinating problems of nature. Ultimately, he intends to interest readers who cultivate biological thought from a philosophical point of view. In this period of social upheaval, he adds, "few things are more vital than a clear conception of the scope and workings of natural selection."[12]

CAMOUFLAGE

Thus, thanks to the complementarity of collecting and fieldwork, the theory of natural selection provides an explanation for the adaptive phenomenon of mimicry. Another case, related to camouflage and apparently less complex insofar as it has to do only with camouflage, is often presented in textbooks, popular books, museums, and exhibitions. This is industrial melanism, found in the peppered moth (*Biston betularia*). Until the beginning of the nineteenth century, as evidenced by entomological collections, this moth sported light-colored wings with a peppery appearance, hence its name, *peppered moth*. During the day, when it was perched on the trunk of a birch tree, it could have been confused with the tree's bark, hence the French name, *phalène du bouleau*, or birch moth, and hence also the Latin epithet *betularia*. Starting in 1848, near Manchester, a dark, so-called melanic form of *Biston betularia* is seen. Rare for a long time, this form becomes more and more frequent over the decades.

The traditional rarity of the black form could be explained by the fact that on light-colored supports, such as birch trees, predators quickly spotted dark-colored subjects. Conversely, on supports blackened by soot, the black was less visible, was therefore less exposed to predation, and tended to spread. Today, deindustrialization and pollution control measures are likely to change this distribution again.

This explanation, supported by entomological collections, mobilized several generations of researchers and sparked debates and controversies. In a note in the aforementioned work on mimicry among butterflies, Punnett raises the question whether better protection derived from coloration is enough to explain the success of the dark form of *Biston betularia*, or whether this success could not also be explained by greater vigor. Edmund B. Ford in the 1930s and then Bernard Kettlewell in the 1950s introduce an experimental approach to the behavior of moths and predatory birds.[13]

The phenomena of mimicry and camouflage are prime examples for biological philosophy. For instance, René Jeannel, a speleology enthusiast and professor at the French National Museum of Natural History, is known for supporting Alfred Wegener's theory of continental drift with biogeographical arguments. In 1946, in his *Introduction à l'entomologie* (Introduction to entomology), he presents a neo-Lamarckian critique of Darwinian explanations for these phenomena:

> But, in truth, the facts are much more naturally explained if one takes into account the Lamarckian principle of the medium acting on hereditary characters. Just as homochromatic colorations are hereditary reactions of living beings to the luminous excitations perceived by the visual organs, so the bright colors are also complex effects of the actions of light on different organisms under the same conditions of medium. Cases of mimicry are caused by the mimicking creature's lifestyle subjected to the same influences as the mimicked creature.[14]

To Jeannel, the Darwinian explanation of homochromia and mimicry (the bright colorations) is a matter of finalism, and a return to the most outdated aspects of Lamarckism (the inheritance of acquired characters) was the rational approach. The further development of biological knowledge has shown this path to be a dead end. Darwinian theory has become essential in entomology as in other life sciences. Thus, in May 2012, it is possible to read the following on the website of the French Centre National de la Recherche Scientifique (CNRS):

> Mimicry is a widespread phenomenon in nature: many species imitate each other in their appearance in order to better defend themselves from predators. An international consortium involving researchers from the CNRS / MNHN (Laboratory of Origin, Structure and Evolution of Biodiversity) and INRA (Physiology of the Insect: Communication and Signaling) has for the first time assembled the complete genome of

the tropical butterfly *Heliconius melpomene*. With this reference genome, researchers show that mimetic similarity is made possible through the exchange of color genes between different species. Until now, hybridizations between neighboring species have been seen as harmful because they produce offspring that are generally less competitive and perform less well. In reality, they also allow the transfer of genes offering a selective advantage, here the colored mark that these butterflies are toxic for their predators.

The most recent molecular biology and sophisticated lab methods partially confirm and partially correct the descriptions and explanations established by fieldwork and the use of collections. Complementarity between experiments and observations also show up in genetics, albeit in a less obvious way.[15]

FRUIT FLIES AND GENETICS

The fate of the Austrian monk Johann Gregor Mendel has aroused the curiosity of several historians of science who question the traditional image of a clergyman tracking pea plants in a Moravian monastery and thus discovering the principles of genetics thirty-five years before the rest of the world. This image is not false, but rather incomplete.[16] Mendel's work needs to be placed in the context of plant hybridization for horticultural and agronomic use. Opportunely, thirty-five years later, Hugo de Vries, Carl Correns, and Erich Tschermak rediscover genetics, thus avoiding a priority dispute. Genetics was thus founded twice—by Mendel in 1865 and by those who rediscovered it in 1900. Mendel's laws are initially established for plants and then applied to animals from the first years of the twentieth century.

Lucien Cuenot, in Nancy, experiments on mice and demonstrates that their pigmentation obeyed Mendel's laws. However, the most used animal model in this field is unquestionably the fruit fly, *Drosophila melanogaster*. As Michel Morange notes in his *Histoire de la biologie moléculaire* (A history of molecular biology), the expansion of genetics is linked to the choice of this species, which was made by the American biologist Thomas Hunt Morgan.[17] This choice proves particularly fortunate: in addition to the low cost of breeding, this insect breeds very quickly, it has only four pairs of chromosomes, and its salivary glands contain giant chromosomes.

In 1915 Morgan publishes a book called *The Mechanism of Mendelian Heredity* with three of his students (Alfred Henry Sturtevant, Hermann Joseph Muller, and Calvin Blackman Bridges). The authors point out that in 1865 the principle of independent segregation of characters was not known. This translates into the fact that in Mendel's pea plants the offspring could have the color of one of the grandparents and the size of another. Morgan considered the chromosomal theory to be a good explanation for the mechanism of heredity. He states that for chromosomes to be considered as the material carriers of heredity, all factors carried by the same chromosome would need to remain together. And this is observed in fruit flies. For example, we can explain the inheritance of eye color in fruit flies by assuming, on the one hand, that the red-eye character is dominant while the white-eye character is recessive and, on the other hand, by admitting that this character is controlled by a gene located on the same chromosome as that which determines sex.

The social and intellectual history of Morgan and his associates is recounted and analyzed by Robert E. Kohler in a book whose title, *Lords of the Fly*, is a nod to William Golding's novel *The Lord of the Flies*, published forty years earlier.[18]

Morgan, who won the Nobel Prize for medicine and physiology in 1933, initially works at Columbia University in New York. In 1928, he accepts a position at Caltech in Pasadena, California. There, Theodosius Grygorovych Dobzhansky, a Ukrainian naturalist who emigrated to the United States, joins the team. He is more specifically interested in variations within species. For this, he focuses on studying another species of the genus Drosophila (*Drosophila pseudoobscura*). Dobzhansky encounters this species while doing research on the sterility in interspecific hybrids.[19] He travels to the Rocky Mountains to capture specimens and map the distribution of natural populations.[20] As both an entomologist and geneticist, with a lot of field experience, he marks the history of biology as one of the founders of the modern synthesis. He is the author of the famous formula "Nothing in biology makes sense except in the light of evolution."[21] Driven by this conviction, he extends his analyses to the problems raised by a biological approach to human societies. This is the subject of an essay, entitled "Heredity and the Nature of Man," in which he defends the idea that human evolution is simultaneously engaged in two different kinds of evolution, biological evolution and cultural evolution. Dobzhansky views culture, which he defines as the sum of habits, beliefs, customs, language, and techniques in use as a reality peculiar to the human species.[22]

The same problem, with a very different answer, can be found at the heart of the controversies raised by sociobiology.

SOCIOBIOLOGY

Is sociobiology a discipline in its own right or a subdiscipline of ethology or a current of political thought? It is hard to situate sociobiology, which we already mentioned in chapter 4, in reference to political images linked to hives and the anthills. As often is the case in this type of quarrel, querying whether this or that discipline is a science not only raises the question of its scientificity but also that of its uniqueness. Disciplines intersect, entangle, interpenetrate, and subordinate each other. Some even exist prior to having their current name.[23]

From this point of view, sociobiology fits into a three-part story. In the 1930s, with Lorenz, von Frisch, and Tinbergen, ethology is defined as the study of animal behavior. Today the term *ethology* continues to be found, but it has been supplanted by the phrase *behavioral ecology*.[24] Situated chronologically between the two, the word *sociobiology* can be considered as the name under which a large number of studies on animal behavior took place from the 1970s to the 1990s. Some authors lent it a hegemonic vocation that contrasts with the modesty of its starting point. The latter is found in what indeed is a very particular phenomenon: the parthenogenesis of drones. Unfertilized eggs result in male bees (drones) and fertilized eggs in female bees (queens or workers). This is established in the nineteenth century by a Polish priest, Johann (or Jan) Dzierzon, both a theologian and a scientist.[25] As a theologian, Dzierzon is known to have opposed the dogma of pontifical infallibility. As a scientist, he is a pioneer of modern beekeeping.

A century later, in 1964, the British biologist William Hamilton published, in two successive issues of the *Journal of Theoretical Biology*, a paper entitled "The Genetical Evolution of Social Behavior."[26] Hamilton outlined a hypothesis linking the social behavior of Hymenoptera to the number of chromosomes. In ants and bees, females are diploid, which means they have 2n chromosomes, while males are haploid, which means they only have n chromosomes.[27] Hamilton's mathematical demonstration reveals haplodiploidy as explicative. Indeed, a calculation of probabilities establishes that in this case a female can have three-quarters of her genes in common with any of her sisters, but that if she has girls she can only have half her genes in common with her offspring. Thus an ant promotes its genes more surely by nursing sisters than by seeking to reproduce itself. This applies to both ants and bees. However, the fact that queens mate with several males complicates the model and requires additional hypotheses.[28] The real difficulties arise when one postulates that all social behavior in animal species can be linked to genetic facts in a similar way with so direct a causality. Among insects, for example, we cannot apply the explanation by haplo-diploidy to the social behavior of termites. Unlike social Hymenoptera, termites include males and females in each caste. This does not mean there is no connection between termite behavior and their genes, but it does mean that the connection must be established based on other principles. In his article, Hamilton acknowledges that termites have the same relatedness to their siblings as to their potential descendants, but he believes that a "bioeconomic" argument applies to termites and explains why a restriction of fertility in certain individuals (male workers and female workers) is beneficial to all siblings.[29]

All this would no doubt have been confined to rather restricted professional circles without Edward O. Wilson's entry on the scene.[30] A field naturalist, an expert on ants, and cofounder with Robert MacArthur of the biogeographical theory of island populations, Wilson is best known today as a promoter of the term *biodiversity*. His intervention in the field of sociobiology is spectacular.[31] He uses a provocative and somewhat humoristic style, imagining that for a zoologist arriving from another planet all the humanities (in the sense of literature) and all the social sciences would be the part of sociobiology devoted to *Homo sapiens*.[32] Perhaps the time has come, he suggests, to temporarily remove ethics from the hands of philosophers and to biologize it.[33] The following year, in 1976, he thinks he sees that "the division between biology, especially population biology, and social sciences no longer exists."[34] Social scientists perceive this claim, stated as a fact, to be a threat of biology annexing their discipline. They are all the more worried because they suspect any biological theory of social behavior would seek to legitimize socioeconomic inequalities, phenomena of exclusion, and discrimination by naturalizing them. The controversy feeds on institutional rivalries as well as on epistemological cleavages.

In 1985 Presses Universitaires de France publishes a book edited by Patrick Tort, *Misère de la sociobiologie* (Misery of sociobiology). The title is a nod to Marx's book, *Misery of Philosophy*, which itself is a controversial answer to Proudhon's *Philosophy of Misery*. The purpose of Patrick Tort's book is to subject sociobiology to radical criticism: far from being a new discipline, sociobiology is nothing more than social Darwinism under the cover of genetics.

In 1993 *La Fourmi et le sociobiologiste* (The ant and the sociobiologist) is published. The author, Pierre Jaisson, defends sociobiology, which, according to him, has been sentenced without due trial. Four years later, he develops the same theme in an interview published in *La Recherche*. On the other end of the spectrum, Monique Chemillier-Gendreau, a specialist in international law, concludes the review of a book by Wilson with, "We must accept that biology compels us to new questions but certainly not accept the risky answers suggested by sociobiology."[35] Mary B. Campbell, after drawing a lively and scholarly picture of bees in England during John Milton's time, issues warnings about Wilson's studies of ants.[36] For her, these studies, used to serve sociobiology, threaten equality between the sexes, equality between the races, and the rights of homosexuals by appealing to the moral authority of nature. Finally, under the title *La Sociobiology*, Michel Veuille publishes a critical, clear, and precise synthesis, indispensable for anyone interested in the question. Without unnecessary emphasis, he explains how "sociobiology with its bad reputation has been replaced over the years by good behavioral ecology."[37]

Among the theoretical debates raised by sociobiology, two deserve special attention.

The first is the determination of the biological units—genes, individuals, groups, etc.—on which natural selection is carried out. In principle, in Darwinian theory, natural selection retains what is advantageous for the individual. Yet, in *The Descent of Man and Selection in Relation to Sex*,[38] Darwin introduces the idea that behavior could be selected even when not advantageous to the individual, as long as it is beneficial for the group to which this individual belongs.[39] Wilson joins the same line of thought when he integrates group selection in a multilevel selection process. However, not all proponents of sociobiology accept this consensual position. Richard Dawkins, author of the 1976 best seller *The Selfish Gene*, sticks strictly to selection of kin. For him, the individual organism is only a "replicator" and a "vehicle" for the gene. To use the metaphorical language that made Dawkins a media success, an individual's altruism results from the selfishness of his or her genes. Seen from the outside, the difference between Wilson's and Dawkins's ideas may seem very slim. Both fit in a Darwinian framework that they apply without hesitation to human society. Nevertheless, controversy rages between the American naturalist and the British biologist.[40] In this respect, there is no certainty that the concept of a "generous gene," which an American biologist, Joan Roughgarden, develops to substitute for the theory of sexual selection, would allow both positions to be rejected. In any case, Roughgarden broadens the vision of sexual behavior and of behavior simply related to the difference between the sexes. She recalls that in some species of social insects colonies contain multiple queens and cohorts of workers of different lineages. An astonishing example is a "super-colony" in Japan, on the plains of Hokkaido, which has forty-five thousand nests and no fewer than a million queens.[41]

The second point of controversy is naturalization. One could be tempted to think that to characterize behavior as natural corresponds necessarily to justifying it. In fact, this consequence is not evident: a quest for intelligibility does not necessarily imply a quest for legitimation. Supposing some primates, of the species *Homo sapiens*, are discussing whether the breeding of another animal species is morally licit. How could the example of ants and aphids provide the elements of an answer?

Those who reject sociobiology and those who make it a substitute for the humanities implicitly admit a postulate of homogeneity that would link human societies and insect societies, despite all their differences. Without this postulate of homogeneity, why should a humanist be worried about the lessons that the sociobiologist might draw from these observations, and why would a naturalist be so eager to widen the field of sociobiology?

8

WORLDS AND ENVIRONMENTS

Speak, and I will baptize you," Cardinal Polignac says to an orangutan found in a glass cage in the King's Garden. The anecdote in Diderot's *Sequel to the Conversation*, illustrates the disturbing proximity between man and monkey.[1] When Linnaeus created the notion of primate, he translates this proximity into taxonomic terms. The theory of evolution, by giving this proximity a genealogical value, makes it threatening to anthropocentrism, thus causing, in Freud's words, a narcissistic injury.[2] Insects, on the contrary, have a carapace constituting an external skeleton and are organized on a very different principle than mammals. Moreover, because of their small size, contact forces have more influence than gravity does on insect lifestyle. As we cannot imagine architects light enough to land on a flower nor small enough to walk on the ceiling,[3] the geometrical perfection we think we detect in bee constructions and the fertile disorder we concede to ants, easily—but deceptively—bring us back to divine geometry, very close to an "intelligent design."[4] In one way, this drift amounts to making insects the little hands of the Great Architect.[5] Thus, in the bestiary of metaphysics, insects are auxiliaries of theologians, just as monkeys are allies of free thinkers.[6]

AN ORGANIZATIONAL PLAN

The proximity between man and animals is at the heart of late eighteenth- and early nineteenth-century controversies.

In the summer of 1830, Johann Wolfgang von Goethe welcomes a visiting friend speaking with emotion about the "volcano" that had just erupted in Paris. The latter, believing that Goethe was talking about the July revolution that had just overthrown King Charles X, mentions Prime Minister Jules de Polignac and the probable expulsion of the royal family. Goethe interrupts him: what do these people matter to him! He was referring to the debate that broke out at the Academy of Sciences between Étienne Geoffroy Saint-Hilaire and Georges Cuvier.[7] This anecdote is of the kind we love to tell, as it illustrates the interest the poet has in plant and animal morphology, but also because it emphasizes the importance of the notion of an organizational plan. This notion, dear to Geoffroy, is essential when one compares, for example, the arm of a man, the anterior fin of a whale, and the wing of a bat. It throws a wrench into comparative anatomy in attempts to extend the search for a unified structure common to the entire animal kingdom.[8]

Many have seen in this opposition a debate for or against transformism. In fact, the question is whether a vertebrate and an invertebrate—hence a man and an insect—are built from the same plan. That this plan results from shared history is another question. For Cuvier, animals are divided into four large divisions, vertebrates (including man), mollusks, articulates (including insects), and zoophytes. Each of these divisions is characterized by a particular plan, without any possible intermediary plan or any passage from one plan to another.[9] On the other hand, for Étienne Geoffroy Saint-Hilaire, "it seems that nature has enclosed itself within certain limits and has formed all living beings from a single plan, essentially the same in principle, but that it has varied in its accessories."[10]

Is it possible to say that Geoffroy anticipated current biological knowledge? Rereading a past controversy in the light of a recent discovery is an anachronism that many historians of science no longer allow themselves. However, a conscious, regulated, and limited use of anachronism makes it possible to multiply potential theoretical points of view of the same empirical reality and then to compare, contrast, or reconcile these perspectives.

By establishing a relationship between Geoffroy's thought and molecular genetics, today's biology sheds light on the concept of an organizational plan. Edward B. Lewis, Christiane Nüsslein-Volhard, and Eric Wieschaus earned the Nobel Prize for medicine in 1995 for their research in this area.[11] The fruit fly, which played such an important role in Morgan's work with his team, reappears here as a model for a problem linking embryology and genetics. As noted by Hervé Le Guyader in an article published in 2000 in the *Revue d'histoire des sciences*, the discovery in the 1980s of the existence of a complex of homeotic genes common to insects and mammals "caused upheaval in the community of biologists."[12] Among the numerous consequences of this discovery is that it has made it possible to locate an ancestor common to insects and mammals around 550 million years ago. Is it possible to say that, since their appearance, humans share the same world as terrestrial arthropods?[13]

THE STROLLER, THE DOG, AND THE TICK

The notion of world is falsely obvious. André Lalande's *Vocabulaire* (Vocabulary) analyzes the different accepted meanings of the word, from world systems, such as those of Ptolemy and Copernicus, to the world as a social group with its rules and uses, the world of socialities. However, the Baltic-German biologist born in Estonia, Baron Jakob von Uexküll, publishes *A Foray Into the Worlds of Animals and Humans* in 1934 using a different meaning for the term *world*.[14] "Any country dweller who traverses woods and bush with his dog has certainly become acquainted with a little animal who lies in wait on the branches of the bushes for his prey, be it human or animal, in order to dive onto his victim and suck himself full of its blood. In doing so, the one- to two-millimeter-large animal swells to the size of a pea."[15]

Thus setting the scene and unrolling the action, Uexküll presents the actors of this small drama, not without noting that the tick has eight legs, a detail that reminds the reader that it does not belong to the class of insects but to that of arachnids. It is indeed an acarian. "Once the female has copulated, she climbs with her full count of eight legs to the tip of a protruding branch of any shrub in order either to fall onto small mammals who run underneath or to let herself be brushed off the branch by large ones."[16]

The tick orients itself in this ascent to "the tip of a branch" using general sensitivity to light. Blind and deaf, it perceives approaching mammals through the odor of butyric acid given off by sweat glands. This odor "gives the tick the signal" to let go, heading in the direction of the mammal in question. If it falls onto something warm, it then "needs only to use its sense of touch to find a spot as free of hair as possible." Once filled with blood, it falls to the ground and lays its eggs before dying.[17] This description of the life cycle of a tick is reminiscent of the style found in Fabre's *Souvenirs entomologiques* in the way it dramatizes the story in combination with detailed observation.[18]

Uexküll differentiates a purely "physiological" interpretation and a "biological" interpretation in order to expose the reader to the lesson he draws from his observations. It is a surprising use of these terms—is not physiology part of biology?—but it has the merit of clearing out the debate for or against a mechanistic approach to life. "For the physiologist," writes Uexküll, "every living being is an object that is located in his human world."[19] On the contrary, the biologist is convinced that "every living thing is a subject that lives in its own world, of which it is the center." This opposition is coupled with a second opposition: rather than a machine, Uexküll prefers to compare the organism "to the machine operator who guides the machine."[20]

To illustrate this point, Uexküll describes "a flowering meadow in which insects buzz and butterflies flutter," asking the reader to imagine something like a soap bubble around each animal. It represents the animal's environment and it "contained all the features accessible to the subject." If we enter this bubble ourselves, "many qualities of the colorful meadow vanish completely, others lose their coherence with one another, and new connections are created." And Uexküll concludes this thought experiment with, "A new world arises with each bubble."[21] To complement this analysis, Georges Kriszat made drawings representing, for example, a room seen by a dog or by a fly.[22]

Uexküll uses the concepts of world (*Welt*) and milieu or environment-world (*Umwelt*). He applies both to humans and animals. This is precisely what the Dutch anthropologist Frederik Jacobus Johannes Buytendijk contests in a comparative psychology essay published in 1958.[23] While paying tribute to Uexküll, "to whom we owe the axiom 'organisms are subjects and not machines,'" Buytendijk writes, *Man does not have an environment, he has a world*."[24] And he specifies that, in the face of this world, man chooses a point of view and that, even if this choice is not completely free, nonetheless, unlike animals, "man exists by his knowledge and his deeds."[25]

PHENOMENOLOGY AND ZOOLOGY

French philosopher Georges Canguilhem brings the tick back, from a historical epistemology perspective, regarding the evolution of the notion of environment. Canguilhem traces the passage from a mechanical conception of the environment to a biological conception. He demonstrates how we begin with the idea of milieu or environment as a median space and pass to it being like a supporting fluid such as water or air. This is the meaning found in Lamarck, who talks about "circumstances" to designate what we now call environment. With Darwin, milieu takes on the meaning of a vital environment. Canguilhem cites Uexküll's story about the tick to illustrate a biological conception of milieu or environment.[26]

Moreover, Uexküll's analysis, which considers every living being to be "a subject that lives in its own world, of which it is the center," presents a certain convergence with a phenomenological approach, as initiated in Edmund Husserl's work. One of the fundamental intuitions of phenomenology is indeed the correlation between consciousness and the world. Consciousness is always aware of something. The motto of the return to the things in themselves invites a description of the phenomena. This leaves unanswered the question of what reality hides behind these phenomena. The main difficulty arises from the fact that other subjects coexist in my own world. As Husserl writes in *Cartesian Meditations*:

> Experience is original consciousness; and in fact we generally say, in the case of experiencing a man; the other is himself there before us "in person." On the other hand, this being there in person does not keep us from admitting forthwith that, properly speaking, neither the other Ego himself, nor his subjective processes or his

appearances themselves, nor anything else belonging to his own essence, becomes given in our experience originally. If it were, if what belongs to the other's own essence were directly accessible, it would be merely a moment of my own essence, and ultimately he himself and I myself would be the same.[27]

This reasoning emphasizes the role of intersubjectivity but does not render natural science obsolete. Husserl states: "Within the world of men and brutes, we encounter the familiar natural-scientific problems of psychophysical, physiological, and psychological genesis."[28]

The Italian philosopher Giorgio Agamben highlights the philosophical impact of Uexküll's research, pointing out that it is contemporary with quantum physics and avant-garde art in the early twentieth century, and that, like them, marks a break with the anthropocentric universe. Agamben explains: "Where classical science saw a single world that comprised within it all living species hierarchically ordered from the most elementary forms up to the higher organisms, Uexküll instead supposes an infinite variety of perceptual worlds that, though they are uncommunicating and reciprocally exclusive, are all equally perfect and linked together as if in a gigantic musical score."[29]

Agamben then summarizes Martin Heidegger's conception of the world: "The guiding thread of Heidegger's exposition," Agamben writes, "is constituted by the triple thesis: 'the stone is worldless [*weltlos*]; the animal is poor in world [*weltarm*]; man is world-forming [*weltbildend*].'"[30]

In his description of an animal's "poverty in world," Heidegger is inspired by Uexküll, considering his studies to be "'the most fruitful thing that philosophy can adopt from the biology dominant today.'"[31] Heidegger makes many references to Uexküll's work and seems to model some of his descriptions on Uexküll's, as for example in this analysis of insect behavior: "The blade of grass that the beetle crawls up, for example, is not a blade of grass for it at all; it is not something possibly destined to become part of the bundle of hay with which the peasant will feed his cow. The blade of grass is simply a beetle-path on which the beetle specifically seeks beetle-nourishment."[32]

The French phenomenological philosopher Maurice Merleau-Ponty also notes the convergence between phenomenology and Uexküll's research. His course at the Collège de France in 1957–1958 is devoted to the theme of nature, and in it he presents the life cycle of the tick and describes its behavior.[33] He notes with interest this idea from Uexküll: "Us men, we also live in one another's *Umwelt*."[34]

Forty years later, in *A Thousand Plateaus*, whose enigmatic title aims to transcend the image of peaks and to suggest the nonhierarchical composition of the book, Gilles Deleuze and Félix Guattari also tell the story of the tick, referring to Uexküll's research.[35] This common reference found in contexts that are otherwise very different has been noted by several authors, in particular by Élisabeth de Fontenay, in *Le Silence des bêtes* (Silence of the beasts).[36]

The subject is, however, far from exhausted, and the tick paradigm deserves attention. It is part of a reductionist approach that seeks intelligibility of phenomena in the simplicity of causes. Biologists are accustomed to seeing reductionism associated with molecular biology and possibly with cell physiology. What is surprising in Uexküll's tick is that its description is part of the naturalistic tradition: it is based on the reconstitution and interpretation of behavior observed in the field, with lab experimentation only providing complementary information, for example, with an evaluation of the time during which a tick can await its victim.

ETHOLOGY AND ANIMAL ETHICS

Despite the tick's celebrity, its use as a prime example is limited to epistemology and ontology.

However, it is possible to extend it to the field of ethics. It suffices to observe "any country dweller" who finds a tick sucking his dog's blood. Concerned about the animal companion's comfort, he rightly or wrongly fears that the tick carries a pathogenic microorganism. He will remove the tick himself or call on a veterinarian or a more experienced neighbor. Yet this gesture, as beneficial as it may be for the mammal, signs the tick's death sentence. Still, nobody perceives this situation as posing a concerning ethical issue.

One could say of the tick and the dog what the French philosopher Florence Burgat says in an article in the INRA *Courrier de l'environnement* about the mosquito and the pig:

> Compassion—the ability to feel the suffering of others as if it were our own—encompasses the animal world, yet perhaps stops when it reaches the smallest animals with which it is difficult or even impossible to identify, and above all whose experience of suffering and anguish is likely nonexistent. One could not put on equal footing the fact of crushing a mosquito and that of slaughtering a pig, even if one has to wonder what good reasons there are for wanting to crush an insect.[37]

Two arguments are put forward to justify giving less importance to the life of an insect or an arachnid compared to that of a mammal. The first seems to be pure subjectivity. The second tends toward a form of objectivity. In fact, the difficulty of identifying with a small terrestrial arthropod (insect, arachnid, millipede, woodlouse) is a form of thought experiment—the temptation of empathy is hindered by the difference in size and structure—while the refusal to admit that these animals can suffer is based on an intellectual construction involving observations and hypotheses.

The paradox is that the same treatment inflicted on an animal will be morally reprehensible or morally indifferent depending on whether or not the animal in question is capable of suffering. In order to determine this capacity, we call upon anatomical arguments, the structure of their nervous system, and ethology, the peculiarities of their behavior. Thus, to establish whether or not insects suffer, we argue about the relative simplicity of their brain and we highlight the way in which a seriously wounded insect continues its activity as if nothing were wrong—for example, when you cut a bee's abdomen it continues to drink.[38] The question of what constitutes the ethical treatment of insects depends very much on the answer to a scientific question of physiology: this dependence opens a gap in the necessary independence between morality and science.

Moreover, getting rid of an inconvenient creature is all the easier when it does not have the possibility of communicating its potential suffering. To qualify this point of view, we can evoke the discomfort aroused by the unconscious cruelty of children tearing off the legs of a fly or the horror inspired by the process of the Hymenoptera (ichneumons wasps) that paralyzes caterpillars and lays their eggs in them so that when the larvae hatch they can devour the caterpillar alive from the inside out. Darwin uses this behavior as an argument in a letter to Asa Gray criticizing the providentialist vision of the world proposed by natural theology.

> With respect to the theological view of the question; this is always painful to me.— I am bewildered.— I had no intention to write

atheistically. But I own that I cannot see, as plainly as others do, & as I shd wish to do, evidence of design & beneficence on all sides of us. There seems to me too much misery in the world. I cannot persuade myself that a beneficent & omnipotent God would have designedly created the Ichneumonidæ with the express intention of their feeding within the living bodies of caterpillars, or that a cat should play with mice. Not believing this, I see no necessity in the belief that the eye was expressly designed. On the other hand I cannot anyhow be contented to view this wonderful universe & especially the nature of man, & to conclude that everything is the result of brute force. I am inclined to look at everything as resulting from designed laws, with the details, whether good or bad, left to the working out of what we may call chance. Not that this notion at all satisfies me. I feel most deeply that the whole subject is too profound for the human intellect.[39]

It is understandable that the behavior of insects and other small articulated animals makes sense to us, but that the distance between their behavior and ours is generally too great for us to feel more than fleeting empathy. In contrast, vertebrates, and especially warm-blooded vertebrates—birds and mammals— express their emotions, and therefore also their suffering, through a behavioral repertoire that gives us the feeling of con- nivance. This feeling has a biological basis that the French neu- roscientist and philosopher Georges Chapouthier summarizes well: "Thus humans do not fundamentally differ from many of their mammalian cousins, and even from birds, in their emotional relationships with their young, in their emotional expressions of joy or anger, in the mechanics of their socio- sexual relations, which come close with some differences, to that of the monkeys, and in most of their mechanisms of memory."[40]

Jacques Derrida's beautiful description of his cat,[41] and the disturbance he feels being naked in front of the pet, cannot be transposed to a fly or an ant. The author's description of this game of glances highlights the limits of a community of living animals, from which insects and other small terrestrial arthropods are excluded. This exclusion emphasizes that while compassion is the most obvious source of people's ethical attitude toward animals, it cannot suffice, and its application to insects is particularly difficult.

The moment legal terminology was imported into the field of human-animal relationships can be seen as a decisive step.[42]

The Universal Declaration of Animal Rights, proclaimed at Unesco Headquarters in October 1978, subsequently revised and presented to the public in 1990, proposes a list of rights that should be granted to animals. A preamble refers to the common origin of all living things and states that the respect of humans for animals is inseparable from the respect of humans among themselves. Article 1 states that "all animals have equal rights to exist within the context of biological equilibrium," but the following sentence takes the form of a restriction: "This equality of rights does not overshadow the diversity of species and of individuals." Such a sentence is fraught with a contradiction between the affirmation of a respect for "all animal life" and concrete limitations that have meaning only vis-à-vis vertebrates. How does one treat a dead mosquito or ant "with decency" (article 3)? Does the notion of "legal status" (article 9) make sense for a moth or bug?

Defining a series of animal rights has the advantage of providing a formal framework that brings ethical thinking closer to legal formalism. However, the difficulty in this area does not come from insufficient legal technicality, but rather seems related to the fact that the category "animal" encompasses very different beings. Logically, talking about animals is like talking about nonhumans and is as surprising as talking about "non-tigers."[43] In *The Animal That Therefore I Am*, Jacques Derrida writes: "The animal is a word, it is an appellation that men have instituted, a name they have given themselves the right and the authority to give to another living creature," to beings as different as lizards and dogs, protozoans and dolphins, ants and silkworms, all "living things" separated by "infinite spaces."[44]

From there stems the idea of being in a contractual relationship, as partners, on the one hand, nature, and on the other, humanity. Michel Serres invites us to do just that in *The Natural Contract* (1990): "Back to nature, then! That means we must add to the exclusively social contract a natural contract of symbiosis and reciprocity in which our relationship to things would set aside mastery and possession in favor of admiring attention, reciprocity, contemplation, and respect."[45]

Anticipating that he would face objections due to the impossibility of the partners signing such a contract, Serres responds in advance that "the old social contract" also remains "unspoken and unwritten" and that "no one has ever read the original or even a copy."[46] In this myth—because the natural contract is a myth, in the sense of an image that illuminates an obscure question—all natural forms are taken into account, albeit, globally without paying particular attention to one or the other, and therefore without the case of insects being mentioned as such.

Undoubtedly inspired by this approach, but situating it in a resolutely day-to-day context, environmental philosophers Catherine Larrère and Raphaël Larrère propose the concept of a "domestic contract" (1997).[47] The domestic contract is for livestock or pets. As far as insects are concerned, it only applies to bees and, in a less personal way, to silkworms. The habits of mourning hives, mentioned by the researcher Philippe Marchenay, are indicative of the contractual relationship between beekeeper and bees.[48] The implicit contract between man and insect takes the form of aesthetic cocreation in artist Hubert Duprat's work with giant casemakers in the Trichoptera order. Caddisfly larvae live in water and construct protective sheaths around themselves usually made of tiny gravel and twigs from their environment. When they are given gold flakes, semiprecious stones, and gems, their protection takes on a strange beauty.[49]

The contract and the statement of rights appear as foundations from which flow human duties toward animals, as expressed joyfully by Montaigne: "We owe justice to men, and graciousness and benignity to other creatures that are capable of it."[50]

Insects—as well as other terrestrial arthropods—are among the smallest possible beings that remain in the realm of the visible. Insects are also the smallest beings for which we can raise the issues of technique and cognition. Entomologists show us how ants find the shortest path and how bees build cells with fascinating geometry; they explain to us how the collective succeeds where an isolated insect would fail; they reveal to us the nocturnal loves of the giant emperor moth; they invite us to look to hives, anthills, and termite mounds in search of certain phenomena that we allow ourselves to describe as social; they paint for us images of insects driven by the often contradictory impulses of wanting to live and the desire to reproduce.

Neither close to us, as are warm-blooded vertebrates (mammals and birds), nor radically other, as can be plants, insects lend themselves to scientific investigation, literary creation, and philosophical reflection.

NOTES

INTRODUCTION

1. Fontenelle 1825 [1709]:210. See also Poupart 1745 [1704].
2. Darwin 1859:216. All notes referring to the *On the Origin of Species* provide the reference in the original 1859 edition.
3. Lecointre and Le Guyader 2001:318–19.
4. See Lamy 1997.
5. Buffon 1753:92; and Buffon 2007:484.
6. Buffon 1753:91; and Buffon 2007:484.
7. About Réaumur, see Torlais 1961; Drouin 1987; and Drouin 1995.
8. See Drouin 2008.
9. Plato 1979: *Apology of Socrates* 30e.
10. See Sleigh 2003. See also Lhoste and Casevitz-Weulersse 1997.
11. The first, *Les fourmis,* was translated into English as *Empire of the Ants.* See Werber 1991.
12. See Buzzati 1998.
13. See *Microcosmos* (Nuridsany and Pérennou 1996) or *Visages d'insectes* (Vanden Eeckhoudt 1965).
14. Siganos 1985.
15. Bizé 2001.
16. Some of these ideas are mentioned in Drouin 2013.

I. TINY GIANTS

1. Michelet 2011 [1858]:359. Translation: http://www.gutenberg.org/files/44287/44287-h/44287-h.htm#Page_336.

2. Chinery 1974:13.
3. The phasmid in question is an *Acrophylla titan*.
4. Chinery 1974:13.
5. I developed some of the ideas found in this chapter as part of a reflection on the issue of whole and parts in natural systems. See Drouin 2007.
6. The Belgian cartoonist Pierre Culliford, who worked under the pseudonym Peyo, introduced the Smurfs in the magazine *Spirou* at the end of the 1950s.
7. Entomologists currently make a distinction between arachnids and insects but include them among arthropods, along with centipedes and crustaceans. See chapter 2.
8. Pascal 1954, pensée 84 [*347*] ff., pp. 1105–6. Translation: http://www.gutenberg.org/files/18269/18269-h/18269-h.htm.
9. For a discussion about Pascal and mites, see Séméria 1985.
10. Swift 1954 [1726/1735], part 1, chapter 2.
11. This estimation joins that which D'Arcy Thompson reaches based on the giants' steps compared to our own. Thompson 1992 [1917/1961]:29; Swift 1954 [1726/1735]:88, part 2, chapter 1. On D'Arcy Thompson, see passim.
12. Voltaire 1960 [1752].
13. Carroll 1865; see the end of chapter 1, and chapters 2 and 4.
14. Latreille 1798:24.
15. Mulsant 1830:15. On Mulsant, see Perron 2006.
16. Michelet 1858, p. 133. Translation: https://www.gutenberg.org/files/44287/44287-h/44287-h.htm.
17. Hölldobler and Wilson 1996 [1994]:134.
18. Frisch 1969 [1953]. On Karl von Frisch, see his autobiography, Frisch 1987 [1957].
19. Frisch 1964 [1955]:40.
20. Blanchard 1877 [1868]:81.
21. Maeterlinck 1930:207.
22. Delage 1913, not 1912 as erroneously indicated by Cornetz and in *The Life of the Ant*. For more on Delage, see Fischer 1979.
23. Maeterlinck 1930:206–9.
24. Aristotle 1895:236, 7.4.9.

25. Aristotle 1895:236, 10–11.
26. Aristotle 1895:237 12.
27. Haldane 1985 [1927]:5–8.
28. In English in the original.
29. Galileo 1914 [1638]:4.
30. Galileo 1914 [1638]:4.
31. Galileo 1914 [1638]:130–31. See Schmidt-Nielsen 1984:48–49 for a recent take on aquatic animals.
32. Galileo 1914 [1638]:130.
33. Schmidt-Nielsen 1984:42–43. Similarly, see Picq 2003.
34. The references indicating Thompson 1992 refer to the 1961 text (an abridged version of the 1917 text) in its 1992 paperback edition (Cambridge University Press.
35. For a good analysis of this theory, see Bouligand and Lepescheaux 1998.
36. Thompson 1992:15–48.
37. On Georges-Louis Lesage (1724–1803), also spelled Le Sage, and on Pierre Prévost (1751–1839), see Trembley 1987:413, 427.
38. Rameaux and Sarrus 1838–1839; and Rameaux 1858. Also on Rameaux, see Arnould 1975.
39. Thompson 1992:26.
40. Stephen Jay Gould in Thompson 1992:xi.
41. All Cournot citations are from Cournot 1987 [1875]:45–46.
42. On the history of theories of the atom, see Bensaude-Vincent and Stengers 1993:294–303.
43. Thompson 1992:17.
44. Poincaré 1908. See Cornetz 1922.
45. Poincaré 1914:95. For more on Poincaré and epistemology, see Brenner 2003.
46. Schuhl 1947:183.
47. Like mites, spiders are not insects but arachnids.
48. See Dudley 1998.

2. AN INORDINATE FONDNESS FOR BEETLES

1. See Hutchinson 1959:146, note. A similar comment is found in Haldane's *What Is Life?*: "The Creator would appear as endowed with a

passion for stars, on the one hand, and for beetles." Haldane 1949:258; referenced from Pascal Tassy in a personal communication.

2. Lamarck 1914 [1809]:23.

3. Aristotle 1862:124.

4. On the extension of the concept of insect, see Daudin 1926–1927b.1: 288–375.

5. Réaumur 1734–1742:58. *Les Mémoires pour servir à l'histoire des insectes* has six volumes, published from 1734–1742.

6. Réaumur 1734–1742:58.

7. Réaumur 1734–1742:42.

8. Réaumur 1926:46 (in French), 134–35 (in English); Réaumur 1928/1929:14. This long-unpublished text was written around 1742 and was meant to be included in a seventh volume of *Mémoires pour servir à l'histoire des insectes*. A first edition was published in 1926, with an English translation by the American entomologist William Morton Wheeler, who was lecturing in French and allowed to consult the archive thanks to the president of the French Academy of Sciences, Eugène Louis Bouvier (also known as Louis Eugène, and sometimes even Émile Louis), who then published the memoir with Charles Pérez in 1928/1929 (and added a seventh volume to the *Mémoires*). See Réaumur 1926 and Réaumur 1928/1929. See also d'Aguilar 2006:54–56 and 196.

9. Buffon 1749.1:36.

10. See Winsor 1976.

11. Linnaeus 1758 [1744]:337–52. A reprint of the fourth edition (1744) of *Systema naturae* was published in 1758, the same year as the tenth edition of the same work.

12. For a rather old historical retrospective, see chapter 2 of Émile Blanchard's book, Blanchard 1877 [1868]. For a recent and detail overview of the history of entomology in general, see d'Aguilar 2006; as well as Smith, Mittler, and Smith 1973. For a biographical approach to the history of entomology in France, see Lhoste 1987; and Gouillard 2004. To place entomology in the context of the other ecological sciences, see the work of Frank Egerton (2012b). See also several articles published in the *Bulletin of the ESA*: Egerton 2005, 2006, 2008, 2012a, 2013.

13. Chappey 2009.

14. Chappey 2009:258–59.

15. Chappey 2009:258–59. Woodlice now belong among isopod crustaceans.

16. Lamarck 1801:36. Lamarck's works can be accessed online in French: http://www.lamarck.cnrs.fr/. On Lamarck and insects, see Brémond and Lessertisseur 1973.

17. On the classification of spiders, see Canard 2008; as well as Rollard and Tardieu 2011.

18. In fact, *character* is the word used by botanists and zoologists to indicate characteristic points (a set of attributes) in the morphology or behavior of an organism (plant or animal), from which they elaborate classification.

19. Lamarck 1994 [1809]:21.

20. Gillispie 1997.

21. On the theoretical debates during this period, see Daudin 1926–1927a and 1926–1927b; Appel 1987; Corsi 2001; Schmitt 2004; Drouin 2008.

22. Latreille 1810:7–8, 21. On Latreille, see Dupuis 1974. See also Burkhardt 1973.

23. Latreille 1810:12.

24. Chappey 2009:258–59.

25. Cuvier 1989 [1810]:245.

26. Cuvier 1989 [1810]:241.

27. Drouin 1998–1999.

28. Darwin 1859:411–13.

29. Darwin 1859:413.

30. Darwin 1859:413; Darwin doesn't give the reference for this quote, but it can be found in Linnaeus 1966 [1751]:119.

31. A few pages later, Darwin comes back to the Linnaeus quote and explains the role that "an appreciation of many trifling points of resemblance, too slight to be defined" often plays in classification. Darwin 1859:417.

32. Darwin 1859:425.

33. "Have been unconsciously seeking": in Darwin 1859:420.

34. Darwin 1859:455.

35. On Darwin and entomology, see Carton 2011.

36. For an epistemological look at cladism, refer to the introduction of the work by Lecointre and Le Guyader 2001; along with *L'Arbre à remonter*

le temps by Tassy 1991; and the article by Claude Dupuis in the *Cahiers des Naturalistes* (Dupuis 1992). See also Grimaldi 2001.

37. Lévi-Strauss 2002 [2001] (about Lecointre et Le Guyader 2001).

38. Hennig 1987 [1965]:102, http://www.ib.usp.br/hennig/Hennig1965.pdf.

39. Hennig 1987 [1965]:102.

40. Hennig 1987 [1965]:104.

41. Hennig 1987 [1965]:104.

42. Hennig 1987 [1965]:105.

43. Hennig 1987 [1965]:105.

44. On Collembola, see Thibaud 2010. Thysanura are generally unknown to the general public. One species, however, known as silverfish (*Lepisma saccharina*), has been known to be found in our homes.

45. See Cambefort 2010:12–16.

46. See Bitsch 2015. On art and science in entomology, see also Cambefort 2004.

3. AN ENTOMOLOGIST'S POINT OF VIEW

1. Proust 1954 [1921].2:600ff.

2. These naturalist allusions serve the role of signs, whose importance in Proust's work Gilles Deleuze points out. See Deleuze 1970 [1964].

3. Without any other indications being provided by the author, we can infer that Proust was implicitly referring to chapter 6 of book 2 of Michelet's *The Sea*.

4. Massis 1924:61.

5. Gouhier 1983:185.

6. See Pelozuelo 2008.

7. Nodier 1982 [1832]. On the entomologist Nodier, see Magnin 1911.

8. Gould 2002:30–50.

9. Jünger 1969 [1967].

10. On Fabre's life, see Revel 1951; Delange 1989; Cambefort 1999; Tort 2002. For his correspondence with Léon Dufour, see Duris 1991. For his correspondence with the publisher Charles Delagrave, see Cambefort 2002.

11. On the idea of discovery narratives, see Carroy and Richard 1998.

12. Fabre 1925, 7th series, chapter 23; Fabre 1989 [1925].2:425. Translation: https://www.gutenberg.org/files/45812/45812-h/45812-h.htm#chapter-15.

13. Fabre 1989 [1925]:425, 430.

14. Egerton 2013:46–47.

15. Karlson and Lüscher 1959. On the history of the notion of pheromones, see Pain 1988; and Dupont 2002.

16. Cocteau 1963:5.

17. Proust 1954 [1913].1:124. Translated edition: 1922. Trans. C. K. Montcrieff, first published by Chatto and Windus in London and Henry Holt in New York (p. 108 of the 2002 Dover Thrift edition).

18. On Dufour the entomologist, see Duris and Diaz 1987; and Duris 1991.

19. This chapter and the other quoted in the following lines take place in the first series of the *Souvenirs entomologiques*.

20. Fabre 1855:131.

21. Fabre 1855:138–39.

22. Fabre 1855:140.

23. Fabre 1925:202, 2d series, chapter 11; Fabre 1989 [1925].1:424. Translation: https://www.gutenberg.org/files/45812/45812-h/45812-h.htm#chapter-19.

24. Fabre 1925:203, 2d series, chapter 11; Fabre 1989 [1925].1:424.

25. Fabre 1925:256, 3d series, chapter 12; Fabre 1989 [1925].1:648). Translation: http://www.efabre.net/chapter-iv-the-tachytes.

26. See Douzou 1985:101; de Gourmont 1903; and Caillois 1934. See also de Gourmont 1925–1931 [1907].

27. Fabre 1925:330, 5th series, chapter 19; Fabre 1989 [1925].1:1104–5.

28. Fabre 1925:331; Fabre 1989 [1925].1:1105.

29. Fabre 1925:332. Fabre 1989 [1925].1:1106.

30. For recent references, see Judson 2006 [2002].

31. Fabre 1925:317–45, 9th series, chapter 21 and 22; Fabre 1989 [1925].2:832–47.

32. Fabre 1925:331; Fabre 1989 [1925]:839); translation from Beebe 1944:225.

33. Fabre 1925:326; Fabre 1989 [1925]:837; translation from Beebe 1944:231.

34. Lorenço contests the systematic predation of males by females; see Lourenço 2008.

35. See Siganos 1985.

36. Von Frisch 1969 [1953]:230–32; and Chinery 1976 [1973]:307–10. See also Villemant 2005.

37. Fabre 1925:140–41, 8th series, chapter 8; Fabre 1989 [1925].2:528.

38. Fabre 1925:140–41; Fabre 1989 [1925].2:529. Translation: https://www .gutenberg.org/files/45812/45812-h/45812-h.htm#chapter-3/

39. Fabre 1925:13, 1st series, chapter 1; Fabre 1989 [1925].1:130–32. Translation: Fabre 1949:99.

40. For an overview of La Fontaine's life and work, see Népote-Desmarres 1999.

41. Réaumur 1926:59 (in French), 1926:148 (in English); Réaumur 1928/1929:31–32.

42. Hölldobler and Wilson 1996 [1994]:216–17.

43. Moggridge 1873.

44. Fabre 1925:229–43, 5th series, chapter 13; Fabre 1989 [1925].1:1052–60.

45. Fabre 1925, 6th series, chapter 13; Fabre 1989 [1925].2:124–33. Translation: https://archive.org/details/fabresbookofinseoofabr.

46. Fabre 1925, 6th series, chapter 13; Fabre 1989 [1925].2:124).

47. Fabre 1925:140–41 8th series, chapter 8; Fabre 1989 [1925].2:529.

48. For duels, see, for example, Fabre 1925:256, 3d series, chapter 12; Fabre 1989 [1925].1:648); and Fabre 1925:202–3, 2d series, chapter 19; Fabre 1989 [1925].1:424. For praying mantis "love," see Fabre 1925:327–33, 5th series, chapter 19; Fabre 1989 [1925].1:1103–6).

49. Lamore 1969.

50. Fabre 1925:76, 10th series, chapter 5; Fabre 1989 [1925].2:913.

51. Revel 1951.

52. Fabre 1925:243, 6th series, chapter 13; Fabre 1989 [1925].2:132.

53. Fabre 1925, 5th series, chapter 10; Fabre 1989 [1925].1:1027–34.

54. Fabre 1925:14–15, 10th series, chapter 1; Fabre 1989 [1925].2:882.

55. Fabre 1925:244, 6th series, chapter 13; Fabre1989 [1925].2:132. Translation: https://archive.org/stream/fabresbookofinseoofabr/fabresbook ofinseoofabr_djvu.txt.

56. Fabre 1925:262, 3th series, chapter 12; Fabre 1989 [1925].1:651.

57. Compagnon 2001. See also Donald Lamore's stylistic analyses of Fabre's spider texts, which shed light on all of Fabre's work: Lamore 1969.

58. Utamaro 2009 [1788]. On the art of the East, see Lhoste and Henry 1990.

59. Pelozuelo 2007. See also Lestel 2003 [2001]; Cambefort 2010.

60. The book was published in 1964, and the film came out the same year.

61. On amateur entomologist physicians, see Gachelin 2011. On amateur entomologists, see Delaporte 1987 and 1989. On amateur naturalists,

see Bensaude-Vincent and Drouin 1996. On amateurs in general, see Cohen and Drouin 1989.

62. On the profession of entomology, see Didier 2005.

63. See the issue of the journal *Alliage* focusing on amateurs (see *Alliage* 2011). Notably Chansigaud 2011; and Drouin 2011.

64. Mulsant 1830. See Lhoste 1987:66–70. See also Perron 2006 (online).

65. Pinault-Sørensen 1991.

66. Grassé 1962:22, 99, 132, 162, 167. Note that the name Marsilly is also written Marsigli.

67. Sigrist, Barras, and Ratcliff 1999:34–38.

68. See Cambefort 2006. For that matter, the use of the word *promenade* in popularized literature highlights the dimension of leisurely study. See, for example, M.V. O., Anonymous 1855 [1848].

69. See *La libellule et le philosophe*, Cugno 2011.

70. On participatory science, see Julliard 2017; Arpin, Charvolin, and Fortier 2015; and Charvolin 2013.

4. INSECT POLITICS

1. Virgil 2001, *Georgics: Book IV*, trans. A. S. Kline, http://www .poetryintranslation.com/PITBR/Latin/VirgilGeorgicsIV.php. See Albouy 2007 and Raulin-Cerceau 2009:11–18.

2. To compare these representations with current knowledge, see Aron and Passera 2000.

3. Proverbs 6:6–10.

4. For an overview of the history of ideas surrounding bee gender, see Maderspacher 2007.

5. Xenophon 2008 [1949]:69, 7:32–35. The Greek word used here is *hegemon*, "chief," with the feminine article, and not *basilea*, "queen." English translation: http://www.gutenberg.org/files/1173/1173-h/1173-h .htm.

6. For an anthropological approach to Aristotelian conceptions about bees, see the rich and enlightening study by Jean-Pierre Albert (Albert 1989).

7. Aristotle 1907, *The History of Animals*, book 9, chapters 39 and 40. See also book 5, chapters 21 and 22. Translation by D'Arcy Wentworth Thompson, http://classics.mit.edu/Aristotle/history_anim.html.

8. Aristotle 2002, book 3, chapter 10:758b–759b. Translation: Aristotle 1943, https://archive.org/stream/generationofanimooarisuoft/genera tionofanimooarisuoft_djvu.txt.

9. Pliny the Elder 1848–1850, book 11, chapter 4. Translation: http://www .perseus.tufts.edu/hopper/text?doc=Perseus%3Atext%3A1999.02.013 7%3Abook%3D11%3Achapter%3D4.

10. Pliny, the Elder 1848–1850, book 11, chapter 27.

11. Butler 1609, preface.

12. Butler 1609, chapter 4.

13. Butler 1609, chapter 4. On the image of bees and political interpretations in England during the sixteenth and seventeenth centuries, and particularly the topic of gender, one can read two excellent studies: Prete 1991 and Campbell 2006.

14. See chapter 3, this volume, and the beginning of this chapter.

15. Montaigne *Essays*, 1588, book 3, chapter 13. French: page 520 of the Garnier edition, see Montaigne 1962.

16. Réaumur attributes this opinion to Cardan. See Réaumur 1926:74 (in French), 1926:163 (in English). See also Réaumur 1928/1929:49.

17. Cervantes, *Don Quixote*, book 2, chapters 33 and 53. See Drouin 1987.

18. On the history of bee biology, see Caullery 1942.

19. Swammerdam 1758:96; translated from the French citation.

20. Swammerdam 1758:96; translated from the French citation.

21. The term was used by Réaumur, who gave credit to Swammerdam. See Réaumur 1926:73 (in French), 1926:162 (in English); Réaumur 1929/1929:48.

22. Swammerdam 1758:187. Translated from French citation.

23. See Fontenelle 1715 [1686].

24. Mandeville 1990 [1714] (The fable of the bees, or private vices, public benefits); Smith 2009 [1776]:146, book 4.

25. Mornet 1911.

26. Pluche 1732:179.

27. Pluche 1732:180.

28. According to d'Aguilar 2006:54–56 and 196.

29. See chapter 2, this volume.

30. Regarding the Hubers, see Cherix 1989.

31. See Buscaglia 1987:304–5.

32. On slavery see throughout this chapter.

33. The first was a deputy to the convention and authored an education reform. He was assassinated by a royalist for his deciding vote in the death of King Louis XVI. The other brother was a partisan of the political agitator François-Noël Babeuf and had to go into exile (Lhoste 1987:133).

34. Lepeletier de Saint-Fargeau 1836:231.

35. See Smeathman 1786.

36. One can find an update on these classification issues in contemporary scientific works (Jaisson 1993; Passera 1984).

37. Several recent articles, and particularly those by Laure Desutter-Grandcolas, have dealt with the phylogeny of stridulatory apparatus in Orthoptera. See Desutter-Grandcolas and Robillard 2004 (online); Robillard and Desutter 2008 (http://www.scielo.br/scielo.php?pid=S 0001-37652004000200019&script=sci_arttext).

38. See Drouin 1992; and Drouin 2005.

39. On Michelet and entomology, see Jolivet 2007; and Marchal 2007.

40. Michelet 2011 [1858]:375, translation W. H. Davenport Adams, 1875:345 (https://archive.org/details/insect__oomich).

41. In his biography, Éric Fauquet highlights the commercial success of these books; see Fauquet 1990:384–92.

42. A number of studies of Michelet's work have given much attention to his naturalist texts. This is, in particular, the case of Roland Barthes's essay, as it is of the work by Linda Orr and Edward K. Kaplan (Barthes 1954; Orr 1976; Kaplan 1977). Georges Gusdorf even dedicated a whole chapter to this aspect of Michelet's work in his book *Le savoir romantique de la Nature,* in which he presented Michelet as "a *Naturphilosoph* in the Germanic sense of the term" (Gusdorf 1985:278).

43. Michelet 2011 [1858]:xxxix, translation 1875.

44. On Espinas, see La Vergata 1996.

45. Espinas 1977 [1878]:70. On Espinas's ideas, see Feuerhahn 2011; and Brooks 1998.

46. Born in Morges, on Lake Geneva, Auguste Forel was a psychiatrist and an entomologist. See Sartori and Cherix 1983; Pilet 1972. For a fine analysis of the political ambiguities in Forel, see Tort 1996; and Jansen 2001b. See also Lustig 2004.

47. On Maeterlinck, see Bailly 1931; and Gorceix 2005.
48. Duchesne and Macquer 1797:7.
49. Huber 1810:289–314; translation: 1820:546, *Natural History of Ants,* https://archive.org/details/naturalhistoryofoohube.
50. Lacène 1822:25. We know nothing about Antoine Lacène except that he authored two horticultural memoirs, also published in Lyon.
51. Lepeletier de Saint-Fargeau 1836:136.
52. Michelet 2011 [1858]:357–58, translation 1875:333.
53. Huber 1810:307; translation 1820:360.
54. Daubenton in Guyon 2006:424, 429.
55. Latreille 1798:30.
56. Huber 1810:303.
57. Lepeletier de Saint-Fargeau 1836:35 and 340.
58. Huber 1810:309; translation 1820:374.
59. Delille 1808:166.
60. On the application of George Dumezil's conceptions of a tripartite society to feudal imagery, see Le Goff 1964:319–29.
61. Latreille 1798:16.
62. Latreille 1798:17.
63. Huber 1810:301; translation 1820:91.
64. Allusion to a discovery made by Adam Gottlob Schirach in the 1760s, confirmed by François Huber, that in a hive deprived of its queen, queen bees can arise from worker larvae.
65. On Toussenel's anti-Semitism, see Poliakov 1968:383–84.
66. Toussenel 1859:58–66. On Toussenel, see Rigol 2005; and Crossley 1990. See also Roman 2007.
67. See, for example, Passera 1984:88, 96, 114, 240; Jaisson 1993:143; Hölldobler and Wilson 1996 [1994]:146.
68. Huber 1810:308–9; translation 1820:374.
69. Lepeletier de Saint-Fargeau 1836:340.
70. Michelet 2011 [1858]:275.
71. Michelet 2011 [1858]:275–89.
72. Ant slavery is as metaphoric as their civil war, as again it occurs among different species.
73. Huber 1810:210; translation 1820:249.
74. Huber 1810:210; translation 1820:308. According to Forel, an avid Huber reader, the ash-colored ants were *Formica fusca,* the mining ants

Formica rufibarbis, and the Amazons Polyergus rufescens (Forel 1874: 102–3). For more on slavery among ants, see Passera 1984:89 and passim. See also Passera 2006.

75. Huber 1810:258; translation 1820:308.

76. Virey 1819.

77. Lepeletier de Saint-Fargeau 1836:100.

78. Lepeletier de Saint-Fargeau 1836:100.

79. Quatrefages 1854:237–38.

80. His correspondence dated 1856–1857, published in 1998, confirmed the importance Michelet accorded to Pierre Huber's book as he was writing L'insecte (Michelet 1998:379–80).

81. Michelet 2011 [1858]:260; translation 1875:199.

82. Michelet 2011 [1858]:262; translation 1875:200.

83. Michelet 2011 [1858]:272, translation 1875:268.

84. Barthes 1954:35.

85. Berthelot 1886:172–84. On science and philosophy in Marcelin Berthelot's work, see Petit 2007.

86. Berthelot 1897. Berthelot also studied wasps: see Berthelot 1905.

87. See Darwin 1859:219–24. On the English translation of Huber and the reading of Darwin, see Clark 1997.

88. Darwin 1859:220.

89. Fabre 1925:335–56, 6th series, chapter 19; Fabre 1989 [1925].2:177–88.

90. Fabre 1925:335; Fabre 1989 [1925].2:177. Translation: http://ibiblio.org /eldritch/jhf/co2.html.

91. Fabre 1925: 349–50; Fabre 1989 [1925].2:185–86.

92. Darwin 1859:218.

93. Darwin 1871.1:364. On the occurrences of this phrase and overall on the relation between the two naturalists, see Tort 2002:147–68; see also Yavetz 1988 and 1991.

94. Fabre 1925:105, 2d series, chapter 7; Fabre 1989 [1925].1:372.

95. Fabre 1925:105; Fabre 1989 [1925].1:372.

96. Fabre 1925:333, 5th series, chapter 19; Fabre 1989 [1925].1:1106.

97. See Tort 2002:263.

98. Maeterlinck 1963 [1901]:213.

99. Kropotkine 1979:11–20. See also 327–36. The original edition was in English and published in 1902.

100. Maeterlinck 1963 [1901]:233.

101. Amouroux 2007. See Delves Broughton 1927 and 1928.

102. See Petit 1988; Petit 1991; Petit 1999.

103. Bergson 1962 [1907]:173–74, chapter 2.

104. Bergson 1962 [1907]:174. The reference given by Bergson was "Peckham, Wasps, solitary and social, Westminster, 1905, p. 28 and following." A version of this text is available online at the Biodiversity Heritage Library: https://www.biodiversitylibrary.org/bibliography/19371 #/summary.

105. Bergson 1962 [1907]:135, chapter 2.

106. Henri Bergson, *Mind-Energy Lectures and Essays*, in English, trans. Wildon Carr, published by Henry Holt in 1920.

107. Bergson 1964 [1919]:26; translation 1920:33, *Mind-Energy Lectures and Essays*, trans. Wildon Carr (New York: Henry Holt).

108. Bergson 1962 [1932]:283; translation 1963 [1935], *The Two Sources of Morality and Religion*. translated by R. Ashley Audra (New York: Henry Holt).

109. See throughout chapter 7.

110. Ruelland 2004:64.

111. See, for example, Réaumur 1926:99 (in French), 1926:188 (in English); Réaumur 1928/1929: 80–81.

112. Latreille 1798:5.

113. Dorat-Cubières 1793:19–20. Read at the "Lycée de l'égalité" in 1792.

114. Lalande details the contents of each of the various types of societies based on reading the work of the zoologist Edmond Perrier.

115. Durkheim 1968 [1922]:42–43.

116. See Jansen 2001b.

117. Bouvier 1926:170–71.

118. Favarel 1945:236.

119. Benveniste 1952:1–8; later in Benveniste 1966:56–62. See also Frisch 1987 [1957].

120. Benveniste 1966:59.

5. INDIVIDUAL INSTINCT AND COLLECTIVE INTELLIGENCE

1. Marx 1969 [1867]:139, book 1, chapter 7. Translation: https://www.marxists.org/archive/marx/works/download/pdf/Capital-Volume-I

.pdf, p. 127. The title of François Mitterrand's book *L'abeille et l'architecte* (The bee and the architect) refers to this text by Karl Marx.

2. See the article "Toile d'araignée" (Spider web) on the French Wikipedia site.

3. Other myths give the spider a happier destiny. On this point and many others about spiders, see Rollard and Tardieu 2011:156–61.

4. Fabre 1925, 9th series, chapter 10; Fabre 1989 [1925].2:736. Translation: *The Life of the Spider*, http://www.gutenberg.org/files/1887/1887-h/1887-h.htm.

5. For more on logarithmic spirals, see http://mathworld.wolfram.com/LogarithmicSpiral.html.

6. Fabre 1925, 9th series, chapter 10; Fabre 1989 [1925].2:742.

7. He even called one of his chapters "The Geometry of Insects." Fabre 1925.2:303–18, 8th series, chapter 18; Fabre 1989 [1925]:612–20.

8. Thompson 1992 [1917/1961]:107–25.

9. Darchen 1958. See also Prete 1990.

10. Pappus d'Alexandrie 1982 [1932]:237–39.

11. Ray 1977 [1717]:132–33.

12. Réaumur 1740:398–99, book 5, 8th memoir.

13. Dew 2013.

14. Maraldi 1731 [1712]:307.

15. Koenig 1740:356. For an explanation of this calculation, see Bessière 1963.

16. Koenig 1740:360.

17. Koenig 1740:360.

18. As an example of using final causes in a physics problem, Koenig cited Leibniz's interpretation of the path followed by a light beam that "passes from one medium to another of different density." This problem of the law of refraction, well known in optics, was linked to questions of mechanics in the theoretical arsenal of the violent controversy between Koenig and Maupertuis about the Principle of Least Action. Maupertuis was accused of plagiarism by Koenig, who attributed the discovery of this principle to Leibniz. Koenig argued against a letter Leibniz had written on this subject to Hermann. Maupertuis, in turn, accused Koenig of having made a forgery. The conflict involved Voltaire who, having sided with Koenig, attracted the hostility of Frederick II, favoring Maupertuis. For details of this episode, see Badinter 1999; Radelet de Grave 1998; Bousquet 2013.

19. Fontenelle 1741:35.
20. Huber 1796 [1792]:112.
21. Huber 1796 [1792]:112.
22. Buffon 1753:98.
23. Buffon 1753:98.
24. Buffon 1753:99. For a geometric approach to crystals, see Haüy 1792.
25. Darwin 1859:224–25.
26. Darwin 1859:226–27.
27. Letter from William H. Miller to Charles Darwin dated 14 May 1858, Darwin 1991. On the dodecahedron, see also Haüy 1792.
28.

> The place of ants in Fabre's work remains a great mystery. While they are omnipresent in the south of France, and their great diversity offers an infinite range of behaviors to study, while their very sophisticated social lifestyle could have provided limitless food for thought for Fabre, who loved comparing humans and human societies, ants are practically non-existent as social insects. Would could also note that bees and social wasps were not treated with the interest they could represent for ethologists, sociobiologists and evolutionists.
>
> (GOMEL 2003)

29. Exodus 31:5–6.
30. Lesser 1742.1:348; translation: *Insecto-Theology*, 1799:135, https://archive .org/details/insectotheologyoolyongoog/page/n5.
31. The link between the idea of instinct and a mechanistic vision of the living was analyzed by Charlotte Sleigh in a chapter called "Ants as Machines" (Sleigh 2004).
32. Darwin 1859:207–8.
33. Darwin 1859:209.
34. Darwin 1859:209.
35. Fabre 1925, 6th series, chapter 18; Fabre 1989 [1925].2:40.
36. Marais 1950 [1938]:90.
37. Maeterlinck 1927 [1926]:144.

38. While the idea was old, the terminology was new. For more on the history of this analogy before Wheeler, see Perru 2003. On organicist philosophy, see Schlanger 1971.

39. Wheeler 1928:304. For a very enlightening historical analysis, see Theraulaz and Bonabeau 1999.

40. On Wilson, see chapter 7.

41. Wilson 1984.

42. Chauvin 1974:71.

43. On pheromones, see chapter 3, this volume.

44. Grassé 1959:65, cited in Theraulaz and Bonabeau 1999:102.

45. Bonabeau and Theraulaz 2000:68. See also Becker et al. 1989.

46. Gordon 1996. On the controversy between Deborah M. Gordon and Edward O. Wilson, Charlotte Sleigh's analysis is useful: Sleigh 2003:167–91.

47. On swarm intelligence, see Miller 2007.

48. See Deneubourg et al. 1991.

49. Bonabeau et Theraulaz 2000:69.

50. See Alaya, Solnon, and Ghedira 2007 [2005]. As explained in the abstract, "The objective is to select a subset of objects that maximize a given utility function, while respecting certain resource constraints."

51. Gordon 2007.

52. Atlan 2011:176.

53. Atlan 2011:78.

6. BATTLES AND ALLIANCES

1. Blanchard 1877 [1868]:14–15. See also Fabre 1922 [1873].

2. On forensic entomology, see Gaudry 2010; and Benecke 2001.

3. Bachelard 1969 [1938]:199–200.

4. Descartes 1953 [1641]:279, Méditation seconde. Translation: *The Method, Meditations, and Philosophy of Descartes,* 1901:230, trans. John Veitch, https://oll.libertyfund.org/titles/descartes-the-method-meditions -and-philosophy-of-descartes.

5. Descartes 1953 [1641]:280; 231.

6. On the history of beekeeping, see Gould and Gould 1993 [1988]:8–25. Bees have also been called "Mouches à miel" (honey flies), which

explains the humorous title that Pierre Déom and Christine Déom gave to a double issue of *La Hulotte* that was dedicated to bees (2010, nos. 28–29). See Déom 2010 [1975].

7. See Favier 1991. See also Huyghe and Huyghe 2006:157–83.

8. Serres 2001 [1600]:713.

9. Serres 2001 [1600]:710–70.

10. On Henri IV's France, see Duby 1971.2, chapter 3 and 4, especially pp. 98–110, and pp. 122–130.

11. Fabre 1925:347–48, 9th series, chapter 23.

12. Perrin 2008, 2009; Sleigh 2003; Cambefort 1994.

13. Barataud 2004 gives and overview of recent research, and in particular that of Roland Lupoli. See Lupoli 2011.

14. Exodus 10:12–15.

15. See the introduction to the Exodus in *La Bible* 1973:135–43. For identification of the insects, see Courtin 2005a and Courtin 2005b. See also Albouy 2006.

16. See Carton et al. 2007.

17. See all of the contributions in the issue of *Parassitologia* dedicated to "Insects and Illness" (Coluzzi et al. 2008).

18. Quinquina, from which Pierre Pelletier and Joseph Caventou extracted quinine in 1820, has long been the only known cure for malaria.

19. Ross 1902. On the mathematization of this area, see Mandal, Sarkar, and Somdata 2011. See also Smith et al. 2012.

20. Latour 1984:127–30.

21. Ghosh 2008 [1996].

22. See Delaporte 2008.

23. Ross 1902.

24. Lotka 1925:81–83. See also Israel and Millán Gasca 2002.

25. Volterra and d'Ancona 1935:10–11.

26. See Kingsland 1985.

27. D'Aguilar, in Robert 2001 [1936]:26–27.

28. Jourdheuil, Grison, and Fraval 1991.

29. See Jansen 2001a.

30. About Haber, see the article by Bretislav (2005–2006). See also "Fritz Haber," in Wikipedia. To situate Haber in the history of chemistry, see Bensaude-Vincent and Stengers 1993:229, 230, 247. See also Vandermeulen 2005, 2007, 2010, 2014.

31. See Aguilar 2006:143.

32. Dajoz 1963:133–35.

33. On Riley, see Acot 1981a and 1981b; Acot 1998:160–62. See also Egerton 2013.

34. See Jourdheuil, Grison, and Fraval 1991:39.

35. Carton et al. 2007.

36. See Perrin 2010.

37. See Jolivet 1991.

38. Théophraste 2003, vol. 1, book 2, chapter 8, § 4.

39. Camerarius 1694.

40. See Hoquet 2005.

41. Linneus 1966 [1751], chapter 5, "Sexus."

42. Dobbs 1750. Dobbs used the word *farina* instead of *pollen*. Both words stem back to the word for flour in Latin.

43. On Philip Miller, see Magnin-Gonze 2009 [2004]; Elliott 2011.

44. Miller (Philip) 1759.

45. Sprengel 1793. See King 1975. See also Magnin-Gonze 2009 [2004]:160.

46. Gedner [and Linnaeus], "What Is It For? (1752) in Linnaeus 1972: 149–50. Linnaeus is considered to be the true author of his students' dissertations (Stafleu 1971:143–55).

47. It can be found in the third volume of *Amoenitates Academicae* and was republished by Camille Limoges with four other Linnaean texts that work together to give an overall vision of nature Linnaeus 1972:145 (note).

48. How natural theology instrumentalized entomology can be found in Kirby and Spence 1814.

49. For more on this theme, see Drouin 1993 [1991].

50. Cugno 2011.

51. Barbault and Weber 2010:48.

52. See Valk 2007.

53. For an objective evaluation, see Barbault and Weber 2010:47–49.

54. Natura 2000 is a European ecological network for which the member states commit to maintaining areas of natural habitats in a suitable state of conservation.

55. See the site http://www.patrickblandin.com/fr/conservation-de- la -nature.

56. See http://www.patrickblandin.com/fr/conservation-de- la-nature.

57. Patrick Blandin, personal communication.
58. See the report of 24 October 2002 session in the *Journal officiel,* dated 25 October 2002. Online: http://www. assemblee-nationale.fr/12/cri /2002-2003/20030034.asp.
59. See Beurois 2001.

7. MODEL INSECTS

1. Borges 1989:225. Translation: Borges 1999.
2. Rousseau 1969 [1762]:772, book 5. Translation: Rousseau 1921.
3. Rousseau 1969 [1762]:772.
4. Bernardin de Saint Pierre 1840 [1784]:137–38. On Bernardin de Saint-Pierre, Rousseau and natural history collections, see Drouin 2001.
5. D'Aguilar, in Robert 2001 [1936]:25.
6. Bates 1862:502.
7. Bates 1862:514–515.
8. See Drouin and Lenay 1990:63–99.
9. Wallace 1897 [1889]:239–240. See Egerton 2012a and Egerton 2012b:170.
10. Darwin cited in Carton 2011:127.
11. Fischer and Henrotte 1998.
12. Punnett 1915:v–vi.
13. Jean Gayon provides a clear and concise presentation of this history (Gayon 1992:373–75). See also Cook 2003.
14. Jeannel 1946:62.
15. For a recent and scholarly overview of the conceptual history of genetics, see Deutsch 2012. On the fruit fly as a model, see Galperin 2006; and Gayon 2006.
16. Drouin 1989.
17. Morange 1994:24.
18. Golding 1954; Kohler 1994. On the same topic as Kohler, see also Bousquet 2003.
19. Bousquet 2003:21–51; Kohler 1994.
20. See Bousquet 2003:40; Kohler 1994:262–93.
21. Dobzhansky 1969 [1965]:120.
22. Dobzhansky 1969 [1965]:145.
23. See Ratcliff 1996.

24. Dominique Vitale, personal communication.
25. See French Wikipedia post on Dzierzon, consulted 6 April 2013. English Wikipedia post consulted 14 March 2018.
26. Hamilton 1964. Hamilton's hypothesis, see Coco 2007.
27. It is known that males and females in the majority of living species have the same number of chromosomes.
28. See Veuille 1997 [1986]:47.
29. See Hamilton 1964.
30. On Wilson, see Sleigh 2003.
31. According to Wilson 1978, John P. Scott created the term *sociobiology* in 1946, and Charles F. Hocket used it in 1948. It was used sporadically between 1950 and 1970.
32. Wilson 1975:547.
33. Wilson 1975:562.
34. Wilson 1976:217.
35. Chemillier-Gendreau 2001:12.
36. Campbell 2006. About the seminare, "Moral Authority of Nature," see Daston and Vidal 2004.
37. Veuille 1997 [1986]:123.
38. Patrick Tort provides a demonstration of how this title was improperly translated for a long time into French. Tort changes *descendance* to *filiation*. See Darwin 1999.
39. Darwin 1871. In the 1999 French edition, Patrick Tort develops the top of the reversive effect of evolution (Darwin 1999).
40. Thorpe 2012.
41. Roughgarden 2012:18.

8. WORLDS AND ENVIRONMENTS

1. Diderot 1964 [1769]:384–85. See Fontenay 1998:329.
2. See Freud 1920 [1917].
3. On insects walking on ceilings, see Guillaume 2001.
4. An individualized observation of ants shows them a lot less unanimous in ardor at work. See Lestel 1985.
5. For an analysis of this trend in Fabre, see Tort 2002.
6. See Séméria 1985.

7. Lacoste 1997:68.

8. On Geoffroy Saint-Hilaire's thought, see Fischer 1999.

9. Lacoste 1997:68.

10. Geoffroy 1796:20, cited in Geoffroy 1818:408–9.

11. Cited and explained by Catherine Bousquet (Bousquet 2003:45).

12. Le Guyader 2000:377.

13. Le Guyader 2000:377.

14. On Uexküll's life, politics, and particularly his nationalism, see Rüting 2004.

15. Uexküll 1965 [1934]:16. Translation: Uexküll 2010:44.

16. Uexküll 1965 [1934]:16.

17. Uexküll 1965 [1934]:16.

18. Uexküll knew Fabre's work. See, in particular, his reference to Fabre's observations of sexual attraction in emperor moths (Uexküll 1965 [1934]:49, translation page 87). See chapter 3, this volume.

19. Uexküll 1965 [1934]:17; translation, p. 45.

20. Uexküll 1965 [1934]:17; translation, p. 45.

21. Uexküll 1965 [1934]:14; translation, p. 43.

22. Uexküll 1965 [1934]: plates between pp. 80, 81; translation, p. 43, plates.

23. Buytendijk 1965 [1958]:54.

24. Buytendijk 1965 [1958]:56 (original emphasis).

25. Buytendijk 1965 [1958]:56–57.

26. Canguilhem 2009 [1965]:184–86.

27. Husserl 1966 [1931]:97; translation: trans. Dorion Cairns, Martinus Nijhoff, https://archive.org/details/CartesiamMeditations.

28. Husserl 1966 [1931]:120.

29. Agamben 2004 [2002]:40.

30. Agamben 2004 [2002]:51.

31. Agamben 2004 [2002]:51.

32. Heidegger 1995 [1983]:198. Heidegger's support of the Nazi Party has been the topic of much controversy. However, it does not appear that his analyses of the animal world were tainted by his politics. On Heidegger and animality, see Elisabeth de Fontenay's chapter on the topic (Fontenay 1998:661–75). See also Pieron 2010.

33. Merleau-Ponty 1995:228–34. On Merleau-Ponty and Uexküll, see Ostachuk 2013.

34. Merleau-Ponty 1995:232.

35. Deleuze and Guattari 1980; Deleuze and Guattari 1987 [1980]. See also *A comme animal*, the first episode of the serial film *Abécédaire* (2004). *Abécedaire* is an eight-hour documentary in which the French philosopher Gilles Deleuze approaches an alphabetical list of words from a philosophical point of view. It was shot in 1988–1989. The first episode is titled *A for Animal*. Among other animals, he mentions the tick. This movie was broadcast episodically on the French television station Arte in 1995 and exists in DVD format (2004).

36. See Fontenay 1998. See also Bailly 2007:87; Goetz 2007; Buchanan 2008.

37. Burgat 2001:66. For Florence Burgat's thoughts on animality, see specifically Burgat 2002; and Burgat 2004.

38. Uexküll, *Theoretische Biologie*, p. 141, cited in Jollivet and Romano 2009:292–93.

39. Darwin, letter to Asa Gray, 22 May 1860, https://www.darwinproject.ac.uk/letter/DCP-LETT-2814.xml.

40. Chapouthier 2004:108.

41. Derrida 2006:20–21.

42. The texts that were collected, presented, and translated by Hicham-Stéphane Afeissa and Jean-Baptiste Jeangène Vilmer are a precious source on animal ethics. See Afeissa and Jeangène Vilmer 2010. See also Bergandi 2013.

43. Drouin 2000:58.

44. Derrida 2006:54–57; translation in Derrida 2002.

45. Serres 1990:67; translation in Serres 1995 [1992]:38.

46. Serres 1990:69; translation in Serres 1995 [1992]:39.

47. Larrère and Larrère 1997.

48. Marchenay and Berard 2007:52, note.

49. See http://trichoptere.hubert-duprat.com/.

50. Montaigne 1962.1:478, book 2, chapter 11 (cited in Larrère and Larrère 1997).

BIBLIOGRAPHY

Acot, Pascal. 1981a. "L'Histoire de la lutte biologique. Première partie: Des origines à la découverte du pouvoir insecticide du DDT." *Courrier de la nature*, no. 75 (September-October): 2–8.

——. 1981b. "L'Histoire de la lutte biologique. Deuxième partie: De la découverte des nouveaux insecticides, du DDT à la lutte intégrée." *Courrier de la nature*, no. 76 (November-December): 8–12.

——, ed. 1998. "The Structuring of Communities." *The European Origins of Scientific Ecology*, 1:151–65. Trans. B. P. Hamm. Amsterdam: Éditions des Archives Contemporaines.

Afeissa, Hicham-Stéphane, and Jeangène Vilmer, Jean-Baptiste, eds. 2010. *Philosophie animale: Différence, responsabilité et communauté.* Paris: Vrin.

Agamben, Giorgio. 2004 [2002]. *The Open: Man and Animal.* Ed. Werner Hamacher. Trans. Kevin Attell. Stanford: Stanford University Press.

——. 2006 [2002]. *L'ouvert. De l'homme et de l'animal.* Trans. Joël Gayraud. Paris: Rivages.

Aguilar, Jacques d'. 2001. "Preface." In Paul-André Robert, *Les Insectes.* Revised ed. Jacques d'Aguilar. Paris: Delachaux et Niestlé.

——. 2006. *Histoire de l'entomologie.* Paris: Delachaux et Niestlé.

——. 2008. "Jan Swammerdam, ou le génie envoûté." *Insectes*, no. 151: 23–24.

——. 2011. "Riley ou une trop éclatante réussite." *Insectes*, no. 160: 23–24.

Alaya, Inès, Christine Solnon, and Khaled Ghedira. 2007 [2005]. "Optimisation par colonies de fourmis pour le problème du sac à dos multi-dimensionnel." *Techniques et Sciences Informatiques* 26, nos. 3–4: 371–90. http://liris.cnrs.fr/csolnon/publications/TSI2006.pdf.

Albert, Jean-Pierre. 1989. "La Ruche d'Aristote: Science, philosophie, mythologie." *L'Homme*, 29, no. 110: 94–116.

Albouy, Vincent. 2006. "Sauterelles ou Éphémères? De la lettre du texte à la réalité quotidienne." *Insectes*, no. 146, pp. 41–42.

——. 2007. "La génération spontanée des Abeilles: Fable paysanne ou mythe érudit?" *Insectes*, no. 145: 22.

Alliage. 2011. Special issue "Amateurs?," no. 69.

Amouroux, Rémy. 2007. "De l'entomologie à la psychanalyse." *Gesnerus* 64: 219–30.

Anonymous. 1855 [1848]. *Promenades d'un naturaliste par M.V.O*, 3d ed. Tours: Mame. Contains "Promenade entomologique, ou Entretien sur les particularités les plus remarquables de l'histoire naturelle des insectes," pp. 141–232.

Anonymous. 2011. "The Evolution of Honeycomb." In Jim Secord, ed., *The Darwin Correspondence Project*. http://www.darwinproject.ac.uk/the -evolution-of-honey-comb.

Appel, Toby. 1987. *The Cuvier-Geoffroy Debate: French Biology in the Decades Before Darwin*. Oxford: Oxford University Press.

Aristotle. 1862. *History of Animals*. Trans. Richard Cresswell. London: Henry G. Bonh.

——. 1895. *Aristotle's Politics: A Treatise on Government*. Trans. William Ellis. London: Routledge.

——. 1907. *The History of Animals*. Trans. D'Arcy Wentworth Thompson. http://classics.mit.edu/Aristotle/history_anim.html.

——. 1943. *Generation of Animals*. Trans. A. L. Peck. Cambridge, MA: Harvard University Press.

Arnould, Pierre. 1975. "Les sciences physiologiques et physico-chimiques." Special centenary issue (1874–1974), *Revue des Annales Médicales de Nancy*. http://www.professeurs-medecine-nancy.fr/Rameaux_J.htm.

Aron, Serge, and Luc Passera. 2000. *Les sociétés animales: Évolution de la coopération et de l'organisation sociale*. Brussels: De Boeck Université.

Arpin, Isabelle, Florian Charvolin, and Agnès Fortier. 2015. "Les inventaires naturalistes: Des pratiques aux modes de gouvernement." *Études rurales* 195:11–16.

Atlan, Henri. 2011. *Le vivant post-génomique, ou Qu'est-ce que l'autoorganisation?* Paris: Odile Jacob.

Bachelard, Gaston. 1969 [1938]. *La Formation de l'esprit scientifique.* Paris: Vrin.

Bacon, Francis. 2001 [1620]. *Novum organum.* Paris: PUF.

Badinter, Élisabeth. 1999. *Les passions intellectuelles*, I, *Désirs de gloire*, 1735–1751. Paris: Fayard.

Bailly, Auguste. 1931. *Maeterlinck.* Paris: Firmin-Didot.

Bailly, Jean-Christophe. 2007. *Le Versant animal.* Paris: Bayard.

Barataud, Bérangère. 2004. "Des insectes comme nouvelle source de médicaments." *Insectes*, no. 132: 29–32.

Barbault, Robert, and Jacques Weber. 2010. *La vie, quelle entreprise! Pour une révolution écologique de l'économie.* Paris: Seuil.

Barthes, Roland. 1954. *Michelet.* Paris: Seuil.

Bates, Henry Walter. 1862. "Contributions to an Insect Fauna of the Amazon Valley." *Transactions of the Linnean Society of London* 23:495–566.

Becker, R., S. Goss, J.-L. Deneubourg, J.-M. Pasteels. 1989. "Colony Size, Communication, and Ant Foraging Strategy." *Psyche* 96, nos. 3–4: 239–56.

Beebe, William, ed. 1944. *The Book of Naturalists: An Anthology of the Best Natural History.* Princeton: Princeton University Press.

Benecke, Mark. 2001. "A Brief History of Forensic Entomology." *Forensic Science International* 120:2–14.

Bensaude-Vincent, Bernadette, and Isabelle Stengers. 1993. *Histoire de la chimie.* Paris: La Découverte.

Bensaude-Vincent, Bernadette, and Jean-Marc Drouin. 1996. "Nature for the People." In Nick Jardine, Jim Secord, and Emma Spary, eds., *Cultures of Natural History*, pp. 408–25. Cambridge: Cambridge University Press.

Benveniste, Émile. 1952. "Communication animale et langage humain." *Diogène* 1:1–8.

——. 1966. *Problèmes de linguistique générale.* Paris: Gallimard.

Bergandi, Donato, ed. 2013. *The Structural Links Between Ecology, Evolution, and Ethics: The Virtuous Epistemic Circle.* Dordrecht: Springer.

Bergson, Henri. 1962 [1907]. *L'évolution créatrice.* Paris: PUF.

——. 1962 [1932]. *Les deux sources de la morale et de la religion.* Paris, PUF. English edition: 1963 [1935]. *The Two Sources of Morality and Religion.* Trans. R. Ashley Audra. New York: Henry Holt.

———. 1964 [1919]. *L'énergie spirituelle*. Paris: PUF. English edition: 1920. *Mind-Energy Lectures and Essays*. Trans. Wildon Carr. New York: Henry Holt.

Bernardin de Saint-Pierre, Jacques-Henri. 1840 [1784]. *Les Études de la nature*. In *Oeuvres*. Paris: Ledentu.

Berthelot, Marcelin. 1886. "Les cités animales et leur évolution." In *Science et philosophie*, pp. 172–84. Paris: Calmann-Lévy.

———. 1897. "Les sociétés animales. Les invasions des Fourmis; le potentiel moral." In *Science et morale*, pp. 313–31. Paris: Calmann-Lévy.

———. 1903. "Lettre à monsieur Ludovic Halévy," i–xxxix. In Jules Michelet, *L'Insecte*, Paris: Calmann-Lévy.

———. 1905. "Les insectes pirates. Les cités des Guêpes." In *Science et libre pensée*, pp. 366–401. Paris: Calmann-Lévy.

Bessière, Gustave. 1963. *Le Calcul intégral facile et attrayant*, 2d ed. Paris: Dunod.

Beurois, Christophe. 2001. "La protection de l'entomofaune, un outil du développement durable?" *Insectes*, no. 121: 3–5.

Bible, La. 1973. Ed. and trans. Émile Osty and Joseph Trinquet. Paris: Seuil.

Bitsch, Colette. 2014. "Le Maître du codex Cocharelli: Enlumineur et pionnier dans l'observation des insectes." In Laurence Talairach-Vielmas and Marie Bouchet, eds., *History and Representations of Entomology in Literature and the Arts*. Brussels: Peter Lang.

———. 2015. "Des sciences naturelles avant la lettre: Le surprenant bestiaire des Cocharelli." Thema, Muséum de Toulouse. http://www.museum .toulouse.fr/-/des-sciences-naturelles-avant-la-lettre-le-surprenant -bestiaire-des-cocharelli.

Bizé, Véronique. 2001. "Les 'insectes' dans la tradition orale." *Insectes*, no. 120: 9–12.

Blanchard, Émile. 1877 [1868]. *Métamorphoses, mœurs et instincts des Insectes (Insectes, Myriapodes, Arachnides, Crustacés)*. Paris: Germer Baillière.

Bonabeau, Éric, and Guy Theraulaz. 2000. "L'intelligence en essaim." *Pour la Science*, no. 271: 66–73.

Borges, Jorge Luis. 1989. "Del rigor en la ciencia." In *Obras completas*, vol. 2. Buenos Aires: Emecé.

———. 1999. "On the Rigor of Science." In *Collected Fictions*. Trans. Andrew Hurley. New York: Penguin.

Bouligand, Yves, and Liên Lepescheux. 1998. "La théorie des transformations." *La Recherche*, no. 305: 31–33.

Bousquet, Catherine. 2003. *Bêtes de science*. Paris: Seuil.

——. 2013. *Maupertuis: Corsaire de la pensée*. Paris: Seuil.

Bouvier, Louis Eugène. 1926. *Le communisme chez les insectes*. Paris: Flammarion.

Brémond, Jean, and Jacques Lessertisseur. 1973. "Lamarck et l'entomologie." *Revue d'Histoire des sciences* 26, no. 3: 231–50.

Brenner, Anastasios. 2003. *Les origines françaises de la philosophie des sciences*. Paris: PUF.

Bretislav, Friedrich. 2005, 2007, 2010, 2014. "Fritz Haber (1868–1934)." http://www.fhi-berlin.mpg.de/history/Friedrich_HaberArticle. pdf.

Brooks III, John I., 1998. "The Eclectic Legacy." *Academic Philosophy and the Human Sciences in Nineteenth-Century France*. Newark: University of Delaware Press.

Buchanan, Brett. 2008. *Onto-ethologies, the Animal Environments of Uexküll, Heidegger, Merleau-Ponty, and Deleuze*. Albany: State University of New York Press.

Buffon, Georges-Louis. 1749. *Histoire naturelle*, vol. 1. Paris: Imprimerie Royale. http://www buffon.cnrs.fr. English edition: 1797. *Buffon's Natural History*. Printed for the proprietor. https://archive.org/details/buff onsnaturalhio2buff.

——. 1753. *Histoire naturelle*, vol. 4. Paris: Imprimerie Royale. http://www.buffon.cnrs.fr. English edition: 1797. *Buffon's Natural History*. Printed for the proprietor. https://archive.org/details/buffonsnatural hio2buff.

——. 2007. *Oeuvres*. Ed. Stéphane Schmitt with Cédric Crémière. Paris: Gallimard.

Burgat, Florence. 2001. "La demande concernant le bien-être animal." *Le Courrier de l'environnement de l'INRA*, no. 44: 65–68.

——. 2002. "La 'dignité de l'animal,' une intrusion dans la métaphysique du propre de l'homme." *L'Homme* 161 (January-March): 197–203.

——. 2004. "Animalité." In *Encyclopedia Universalis*.

Burkhardt, Richard W. 1973. "Latreille, Pierre-André." In Charles C. Gillispie, ed., *Dictionary of Scientific Biography*, 8:49–50. New York: Scribner.

Buscaglia, Marino. 1987. "La zoologie." In Jacques Trembley, ed., *Les Savants genevois dans l'Europe intellectuelle du XVIIe au milieu du 19ᵉ siècle*, pp. 267–328. Geneva: Éditions du Journal de Genève.

Butler, Charles. 1609. *The Feminine Monarchie, or a Treatise Concerning Bees and the Due Ordering of Bees*. Oxford: Oxford University Press.

Buytendijk, Frederik Jacobus Johannes. 1965 [1958]. *L'homme et l'animal*. Trans. Rémi Laureillard. Paris: Gallimard.

Buzzati, Dino. 1998. "Les fourmis." In Charles Ficat, ed., *Histoires de Fourmis*, pp. 7–8. Paris: Les Belles Lettres.

Caillois, Roger. 1934. "La mante religieuse." *Minotaure*, no. 5: 23–26.

Cambefort, Yves. 1994. *Le scarabée et les dieux: Essai sur la signification symbolique et mythique des Coléoptères*. Paris: Boubée.

——. 1999. *L'oeuvre de Jean-Henri Fabre*. Paris: Delagrave.

——, ed. 2002. *Jean-Henri Fabre: Lettres inédites à Charles Delagrave*. Paris: Delagrave.

——. 2004. "Artistes, médecins et curieux aux origines de l'entomologie moderne (1450–1650)." *Bulletin d'Histoire et d'Épistémologie des Sciences de la vie* 11, no. 1: 3–29.

——. 2006. *Des Coléoptères, des collections et des hommes*. Paris: Muséum national d'histoire naturelle.

——. 2007. "Entomologie et mélancolie: Quelques aspects du symbolisme des insectes dans l'art européen du 14ᵉ au 21ᵉ siècle"/"Entomology and Melancholy: Some Aspects of Insect Symbolism in European Art from the Fourteenth to the Twenty-First Century." In Edmond Dounias, Motte Florac Élisabeth, and Dunham Margaret, ed., *Le symbolisme des animaux: L'animal, clef de voûte de la relation entre l' homme et la nature?/ Animal Symbolism: Animals, Keystone in the Relationship Between Man and Nature?*, pp. 393–423. Paris: IRD.

——. 2010. "Des scarabées et des hommes: Histoire des Coléoptères de l'Égypte ancienne à nos jours." In Laurence Talairach-Vielmas and Marie Bouchet, eds., *Spinning Webs of Wonder: Insects and Texts*, pp. 169–208. Explora symposium papers. Toulouse: Publications du Muséum d'histoire naturelle de Toulouse, 4–5 May.

Camerarius. 1694. "Epistola ad D. Mich. Bern. Valentini de sexu plantarum." Tübingen; reprint 1749. Johann Georg Gmelin, *Sermo academicus de novorum vegetabilium* . . . , pp. 83–148. Tübingen; and 1797. Johan Christian Mikan, *Opuscula botanici argumenti*, pp. 43–117. Prague.

Campbell, Mary B. 2006. "Busy Bees. Utopia, Dystopia and the Very Small." *Journal of Medieval and Early Modern Studies* 36:619–42.

Canard, Frédérik, ed. 2008. *Au fil des araignées*. Rennes: Apogée.

Canguilhem, Georges. 2009 [1965]. "Le vivant et son milieu." In *La connaissance de la vie*. Paris: Vrin.

Carroll, Lewis. 1865. *Alice's Adventures in Wonderland*. London: Macmillan.

Carroy, Jacqueline, and Nathalie Richard, eds. 1998. *La Découverte et ses récits en sciences humaines*. Paris: L'Harmattan.

Carson, Rachel. 1962. *Silent Spring*. Boston: Houghton Mifflin.

Carton, Yves. 2011. *Entomologie, Darwin et Darwinisme*. Paris: Hermann.

Carton, Yves, Conner Sørensen, Janet Smith, and Edward Smith. 2007. "Une coopération exemplaire entre entomologistes français et américains pendant la crise du Phylloxéra en France, 1868–1895." *Annales de la société entomologique de France* n.s. 43, no. 1: 103–25.

Caullery, Maurice, ed. 1942. "Développement historique de nos connaissances sur la biologie des abeilles." In *Biologie des abeilles*, pp. 1–26. PUF: Paris.

Chansigaud, Valérie. 2011. "De l'histoire naturelle à l'environnementalisme: Les enjeux de l'amateur." *Alliage*, special issue "Amateurs?," no. 69: 62–70.

Chappey, Jean-Luc. 2009. *Des naturalistes en Révolution: Les procès-verbaux de la Société d'histoire naturelle de Paris, 1790–1798*. Preface by Pietro Corsi. Paris: Éditions du Comité des Travaux Historiques et Scientifiques (CTHS).

Charvolin, Florian. 2013. "Pense-bêtes, astuces et recettes de jardiniers-observateurs de papillons: Retour sur une science citoyenne." *Revue d'anthropologie des connaissances* 7, no. 2: 485–500.

Chapouthier, Georges, ed. 2004. *L'Animal humain: Traits et spécificités*. Paris: L'Harmattan.

Chauvin, Rémy. 1974. "Les sociétés les plus complexes chez les insectes." *Communications* 22:63–71.

Chemillier-Gendreau, Monique. 2001. "Sociobiologie, liberté scientifique, liberté politique: Une critique de Edward Wilson." *Mouvement*, no. 17: 88–98.

Cherix, Daniel. 1990. "De Voltaire aux Fourmis en passant par les Abeilles ou petite chronique de la famille Huber de Genève." *Actes des Colloques Insectes sociaux* 6:1–7.

Chinery, Michael. 1974. *A Field Guide to the Insects of Britain and Northern Europe*. Boston: Houghton Mifflin.

———. 1976 [1973]. *Les insectes d'Europe en couleurs*. Paris: Elsevier Séquoia.

Clark, John Finley McDiarmid. 1997. "'A Little People but Exceedingly Wise?' Taming the Ant and the Savage in Nineteenth-Century England," pp. 65–83. *La Lettre de la Maison Française*, no 7. Oxford: Oxford University Press.

Coco, Emanuele. 2007. *Etologia*. Milan: Giunti.

Cocteau, Jean. 1963. "Lettre de Marcel Proust à Jean Cocteau." *Bulletin de la Société des amis de Marcel Proust et des amis de Combray*, no. 13: 3–5.

Cohen, Yves, and Jean-Marc Drouin, eds. 1989. *Les Amateurs de sciences et de techniques*, Cahiers d'histoire et de philosophie des sciences, no. 27.

Coluzzi, Mario, Gabriel Gachelin, Anne Hardy, and Annick Opinel, eds. 2008. "Insects and Illnesses: Contributions to the History of Medical Entomology." *Parassitologia* 50, nos. 3–4: 157–330.

Compagnon, Antoine. 2001. *Théorie de la littérature: La notion de genre littéraire*. Paris IV Sorbonne. http://www.fabula.org/compagnon/genre .php.

Cook, Laurence. 2003. "The Rise and Fall of the Carbonaria Form of the Peppered Moth." *Quarterly Review of Biology* 76, no. 4: 399–417.

Cornetz, Victor. 1922. "Remy de Gourmont, J.-H. Fabre et les Fourmis." *Mercure de France* 158:27–39.

Corsi, Pietro. 2001. *Lamarck: Genèse et enjeux du transformisme, 1770–1830*. Paris; CNRS.

Cournot, Antoine-Augustin. 1987 [1875]. *Matérialisme. Vitalisme. Rationalisme: Étude sur l'emploi des données de la science en philosophie*. Ed. Claire Salomon-Bayet. Paris: Vrin.

Courtin, Rémi. 2005a. "Insectes et Arthropodes de la Bible, 1re partie." Illustrations Marianne Alexandre. *Insectes*, no. 137: 35–36.

———. 2005b. "Insectes et Arthropodes de la Bible, 2e partie." Illustrations Marianne Alexandre. *Insectes*, no. 138: 34–35.

Crossley, Ceri. 1990. "Toussenel et la femme." *Cahiers Charles Fourier*, no. 1 (December): 51–65. http://www.charlesfourier.fr/article.php3?id_article=7.

Cugno, Alain. 2011. *La libellule et le philosophe*. Paris: l'Iconoclaste.

Cuvier, Georges. 1989 [1810]. *Chimie et sciences de la nature: Rapports à l'Empereur*, vol. 2. Ed. Yves Laissus. Paris: Belin.

Dajoz, Roger. 1963. *Les animaux nuisibles*. Paris: La Farandole.

Darchen, Roger. 1958. "Construction et reconstruction de la cellule des rayons d'Apis mellifera." *Insectes sociaux* 5, no. 4: 357–71.

Darwin, Charles. 1859. *On the Origin of Species*. London: John Murray. 1964. Facsimile of the first edition, introduction by Ernst Mayr. Cambridge, MA: Harvard University Press.

——. 1871. *The Descent of Man and Selection in Relation to Sex*. 2 vols. London: John Murray.

——. 1991. *The Correspondence of Charles Darwin*, vol. 7: 1858–1859. Ed. Frederick Burkhardt, Sydney Smith et al. Cambridge: Cambridge University Press. https://www.darwinproject.ac.uk/letter/DCP-LETT-2814.xml.

——. 1999 [1871]. *La filiation de l'homme et la sélection liée au sexe*. Ed. Patrick Tort. Trans. Michel Prum. Paris: Syllepse.

Daston, Lorraine, and Fernando Vidal, ed. 2004. *The Moral Authority of Nature*. Chicago: Chicago University Press.

Daubenton, Louis Jean-Marie. 2006 [1795]. "Leçons d'histoire naturelle." In Étienne Guyon, ed., *L'École normale de l'an III: Leçons de Physique, de Chimie, d'Histoire naturelle*, pp. 395–572. Paris: Éditions rue d'Ulm.

Daudin, Henri. 1926–1927a [1740–1770]. *De Linné à Lamarck: Méthodes de la classification et l'idée de série en botanique et en zoologie*. 2 vols. Paris: Félix Alcan. Facsimile reprint 1983. Paris: Éditions des Archives contemporaines.

——. 1926–1927b. *Cuvier et Lamarck: Les classes zoologiques et l'idée de série animale, 1790–1830*. Paris: Félix Alcan. Facsimile reprint 1983. Paris: Éditions des Archives contemporaines.

Dawkins, Richard. 1976. *The Selfish Gene*: New York: Oxford University Press.

Delage, Yves. 1913. "La dégradation progressive de la richesse physiologique." *Revue scientifique* 51, no. 3 (19 July): 65–69.

Delange, Yves. 1989. Preface to Jean-Henri Fabre, *Souvenirs entomologiques*, pp. 1–117. Paris: Laffont.

——, ed. 2003. *Jean-Henri Fabre, un autre regard sur l'insecte*. Rodez: Conseil général de l'Aveyron.

Delaporte, François. 2008. "The Discovery of the Vector of Robles Disease." In Mario Coluzi, Gabriel Gachelin, Anne Hardy, and Annick Opinel, eds., *Insects and Illnesses: Contributions to the History of Medical Entomology. Parassitologia* 50, nos. 3–4: 227–31.

Delaporte, Yves. 1987. "Sublaevigatus ou subloevigatus? Les usages sociaux de la nomenclature chez les entomologistes." In Jacques Hainard and

Roland Kaehr, eds., *Des animaux et des hommes*, pp. 187–212. Neuchâtel: Musée d'ethnographie.

———. 1989. "Les entomologistes amateurs: Un statut ambigu." In Yves Cohen and Jean-Marc Drouin, eds., "Les amateurs de sciences et de techniques." *Cahiers d'histoire et de philosophie des sciences*, no. 27: 175–90.

Deleuze, Gilles. 1970 [1964]. *Proust et les signes*. Paris: PUF.

———. 2010. "Uexküll." *Bulletin d' analyse phénoménologique* 6, no. 2: *La nature vivante* (Actes no. 2). http://popups.ulg.ac.be/bap/ document. php?id=384.

Deleuze, Gilles, with Claire Parnet. 2004. *Abécédaire*. Produced by Pierre-André Boutang. Paris: Vidéo éditions Montparnasse.

Deleuze, Gilles, and Félix Guattari. 1980. *Mille Plateaux*. Paris: Éditions de Minuit.

———. 1987 [1980]. *A Thousand Plateaus: Capitalism and Schizophrenia*. Minneapolis: University of Minnesota Press.

Delille, Jacques. 1808. *Les Trois règnes de la Nature*. 2 vols. Paris: H. Nicolle.

Delves Broughton, L. R. 1927. "Vues analytiques sur la vie des Abeilles et des Termites." Trans. Marie Bonaparte. *Revue française de Psychanalyse* 1, no. 3: 562–67.

———. 1928. "Vom Leben der Bienen und Termiten Psychoanalytsche Bermekungen." *Imago*, no. 14: 142–46.

Deneubourg, Jean-Louis, Simon Goss, N. Franks, Ana B. Sendova-Franks, Claire Detrain, and L. Chrétien. 1991. "The Dynamic of Collective Sorting: Robot-like Ants and Ant-like Robots." In J. A. Meyer and S. Wilson, eds., *From Animals to Animats*, pp. 356–65. Cambridge, MA: MIT Press/Bradford.

Deneubourg, Jean-Louis, Jacques Pasteels, and Jean-Claude Verhaeghe. 1984. "Quand l'erreur alimente l'imagination d'une société: Le cas des fourmis." *Nouvelles de la science et des technologie*s 2:47–52.

Déom, Pierre. 2010. *La Hulotte*, special issue "Mouches à miel," nos. 28–29.

Derrida, Jacques.2002. "The Animal That Therefore I Am." *Critical Inquiry* 28, no. 2: 392–402.

———. 2006. *L'Animal que donc je suis*. Ed. Marie-Louise Mallet. Paris: Galilée.

Descartes, René. 1953 [1641]. *Méditations métaphysiques*. In *Oeuvres et lettres*, Paris: Gallimard. English edition: 1901. *The Method, Meditations, and*

Philosophy of Descartes. Trans. John Veitch. https://oll.libertyfund.org /titles/descartes-the-method-meditations-and-philosophy-of-descartes.

Desutter-Grandcolas, Laure, and Tony Robillard. 2004. "Acoustic Evolution in Crickets: Need for Phylogenetic Study and a Reappraisal of Signal Effectiveness." *Anais da Academia Brasileira de Ciências* 76, no. 2 (June 2004). http://www.scielo.br/scielo.php?pid=S0001-376520040002 00019&script=sci_arttext.

Deutsch, Jean. 2012. *Le gène, un concept en evolution*. Preface by Jean Gayon. Paris: Seuil.

Dew, Nicholas. 2013. "The Hive and the Pendulum: Universal Metrology and Baroque Science." In Ofer Gal and Raz Chen-Morris, eds., *Science in the Age of Baroque*, pp. 239–55. Dordrecht: Springer.

Diderot, Denis. 1964 [1769]. *Le Rêqve de d'Alembert*. Paris: Garnier.

Didier, Bruno. 2005. "Métier: entomologiste. Claire Villemant." *Insectes*, no. 138: 23–27.

Dobbs, Arthur. 1750. "A Letter . . . Concerning Bees and their Method of Gathering Wax and Honey." *Philosophical Transactions of the Royal Society* 46:536–49.

Dobzhansky, Théodosius. 1969 [1965]. *L'hérédité et la nature humaine*. Trans. Simone Pasteur. Paris, Flammarion.

Dorat-Cubières, Michel. 1793. *Les Abeilles ou l'Heureux gouvernement*. Paris: Gérod et Tessier.

Douzou, Pierre. 1985. *Les Bricoleurs du septième jour*. Paris: Fayard.

Drouin, Jean-Marc. 1987. "Du terrain au laboratoire: Réaumur et l'histoire des Fourmis." *ASTER, Recherches en didactique des sciences expérimentales*, no. 5: 35–47.

——. 1989. "Mendel, côté jardin." In M. Serres, ed., *Éléments d'histoire des sciences*, pp. 406–21. Paris: Bordas.

——. 1992. "L'image des sociétés d'insectes en France à l'époque de la Révolution." *Revue de Synthèse* 4:333–45.

——. 1993 [1991]. *L'Écologie et son histoire*. Paris: Flammarion.

——. 1995. "Les curiosités d'un physician." In J. Dhombres, ed., *Aventures scientifiques en Poitou-Charentes du 16ᵉ au 20ᵉ siècle*, pp. 196–209. Poitiers: Éditions de l'actualité Poitou-Charentes.

——. 1998–1999. "Les naturalistes: Des révolutionnaires tranquilles." *La Lettre de la Maison française d'Oxford*, no. 10: 132–39.

———. 2000. "Le théâtre de la nature." In Catherine Larrère, ed., *Nature vive*, pp. 48–61. Paris: Nathan et MNHN.

———. 2001. "Rousseau, Bernardin de Saint-Pierre et l'histoire naturelle." *Dix-huitième siècle*, no. 33: 507–16.

———. 2005. "Ants and Bees Between the French and the Darwinian Revolution." *Ludus Vitalis* 12, no. 24: 3–14.

———. 2007. "Quelle dimension pour le vivant?" In Thierry Martin, ed., *Le tout et les parties dans les systèmes naturels*, pp. 107–14. Paris: Vuibert.

———. 2008. *L'herbier des philosophes*. Paris: Seuil.

———. 2011. "Les amateurs d'histoire naturelle: Promenades, collectes, et controverses." *Alliage*, special issue "Amateurs?," no. 69: 35–47.

———. 2013. "Three Philosophical Approaches to Entomology." In Hanne Andersen, Dennis Dieks, Wenceslao J. Gonzalez, Thomas Uebel, and Gregory Wheeler, eds., *New Challenges to Philosophy of Science: The Philosophy of Science in a European Perspective*, pp. 377–86. New York: Springer.

Drouin, Jean-Marc, and Charles Lenay. 1990. *Théories de l'évolution: Une anthologie*. Paris: Presses Pocket.

Duby, Georges, ed. 1971. *Histoire de la France*. 3 vols. Paris: Larousse.

Duchesne, Henri-Gabriel, and Pierre Joseph Macquer. 1797. *Manuel du naturaliste*, vol. 1. 2d ed., Paris: Rémont.

Dudley, Robert. 1998. "Atmospheric Oxygen, Giant Paleozoic Insects, and the Evolution of Aerial Locomotor Performance." *Journal of Experimental Biology*, no. 261: 1043–50.

Dupont, Jean-Claude. 2002. "Les molécules phéromonales: Éléments d'épistémologie historique." *Philosophia Scientiae* 6:100–22.

Dupuis, Claude. 1974. "Pierre-André Latreille (1762–1833): The Foremost Entomologist of His Time." *Annual Review of Entomology* 19:1–13.

———. 1992. "Permanence et actualité de la Systématique: Regards épistémologiques sur la taxinomie cladiste." *Cahiers des Naturalistes* n.s. 48, fasc. 2: 29–53.

Duris, Pascal. 1991. "Quatre lettres inédites de Jean-Henri Fabre à Léon Dufour." *Revue d'histoire des sciences* 44, no. 2: 203–18.

Duris, Pascal, and Elvire Diaz. 1987. *Petite histoire naturelle de la première moitié du XIXᵉ siècle: Léon Dufour, 1780–1865*. Bordeaux: Presses universitaires de Bordeaux.

Durkheim, Émile 1968 [1922]. *Éducation et sociologie*. Paris: PUF.

Dzierzon, Jan. 1884. "L'accouplement récemment observé d'une ouvrière avec un faux bourdon peut-il ébranler ma théorie?," pp. 1–8. Ed. J. B. Leriche. Bordeaux: Imprimerie Durand.

Egerton, Frank N. 2005. "A History of the Ecological Sciences, Part 17: Invertebrates Zoology and Parasitology during the 1600s." *Bulletin of the ESA* 86, no. 3: 133–44. http://www.esajournals.org/loi/ebul.

——. 2006. "A History of the Ecological Sciences, Part 21: Reaumur and the History of Insects." *Bulletin of the ESA* 87, no. 3: 212–24.

——. 2008. "A History of the Ecological Sciences, Part 30: Invertebrate Zoology and Parasitology during the 1700s." *Bulletin of the ESA* 89, no. 4: 407–33.

——. 2012a. "A History of the Ecological Sciences, Part 41: Victorian Naturalists in Amazonia—Wallace, Bates, Spruce." *Bulletin of the ESA* 93, no. 1: 35–59.

——. 2012b. *Roots of Ecology: Antiquity to Haeckel*. Berkeley: University of California Press.

——. 2013. "A History of the Ecological Sciences, Part 45: Ecological Aspects of Entomology During the 1800s." *Bulletin of the ESA* 94, no. 1: 36–88.

Elliott, Brent. 2011. "Philip Miller as a Natural Philosopher." *Occasional Papers from the Royal Society of Horticulture Library* 5:3–48.

Espinas, Alfred. 1977 [1878]. *Des sociétés animales*, 2d ed. Paris: Germer, Baillière et Cie. Facsimile ed. New York: Arno.

Fabre, Jean-Henri. 1855. "Observations sur les mœurs des Cerceris et sur les causes de la longue conservation des Coléoptères dont ils approvisionnent leurs larves." *Annales des sciences naturelles*, 4th series, *Zoologie* 4, no. 3: 129–50.

——. 1922 [1873]. *Les Auxiliaires, récits sur les animaux utiles à l'agriculture*. Paris: Delagrave.

——. 1925. *Souvenirs entomologiques*. Trans. Alexander Teixeira de Mattos. 10 vols. Paris: Delagrave.

——. 1949. *The Insect World of J. Henri Fabre*. Trans. Alexander Teixeira de Mattos. New York: Dodd, Mead.

——. 1989 [1925]. *Souvenirs entomologiques*. Ed. Yves Delange. 2 vols. Paris: Robert Laffont.

Farley, Michael. "L'institutionnalisation de l'entomologie française." *Bulletin de la Société entomologique de France*, no. 88 (1983): 134–43.

Fontenelle, Bernard Le Bovier de. 1825 [1709]. "Éloge de François Poupart." In *Oeuvres*, vol. 1: *Éloges*, pp. 209–12. Paris: Salmon and Peytieux.

Fauquet, Éric. 1990. *Michelet ou la Gloire du professeur d'histoire*. Paris: Le Cerf.

Favarel, Geo. 1945. *Démocratie et dictature chez les Insectes*. Paris: Flammarion.

Favier, Jean. 1991. *Les Grandes découvertes d'Alexandre à Magellan*. Paris: Fayard.

Feuerhahn, Wolf. 2011. "Les 'sociétés animales:' un défi à l'ordre savant." *Romantisme*, no. 154: 35–51.

Fischer, Jean-Louis. 1979. "Yves Delage (1854–1920): L'épigénèse néolamarckienne contre la prédétermination weismannienne." *Revue de Synthèse*, no. 95–96: 443–61.

——. 1999. "Les manuscrits égyptiens d'Étienne Geoffroy Saint-Hilaire." In P. Bret, ed., *L'Expédition d'Égypte, une entreprise des Lumières*, pp. 243–59. Institut de France, Académie des sciences, Tec et Doc Lavoisier.

Fischer, Jean-Louis, and Jean-Georges Henrotte. 1998. "Mimétisme chez les Papillons." *Pour la Science*, no. 251: 56–62.

Fontenay, Élisabeth de. 1998. *Le silence des bêtes: La philosophie à l'épreuve de l'animalité*. Paris: Fayard.

Fontenelle, Bernard Le Bovier de. 1741. *Histoire de l'Académie Royale des sciences pour l'année 1739*, pp. 30–35. Paris: Imprimerie royale.

——. 1825 [1709]. "Éloge de François Poupart." In *Éloges, Oeuvres*, 1:209–12. Paris: Salmon et Peytieux.

——. 1990 [1686]. *Entretiens sur la pluralité des mondes*. Paris, L'Aube. English edition: 1715 [1686]. *Conversations on the Plurality of Worlds*. Trans. William Gardiner. London: Bettesworth.

Forel, Auguste. 1874. *Les fourmis de la Suisse*. Lyon: H. Georg.

Freud, Sigmund. 1920 [1917]. *A General Introduction to Psychoanalysis*. Trans. G. Stanley Hall. New York: Liveright.

Frisch, Karl von. 1959 [1955]. *Dix petits hôtes de nos maisons*. Trans. André Dalcq. Paris: Albin Michel.

——. 1964 [1955]. *Ten Little Housemates*. Trans. M. D. Senft. Oxford: Pergamon.

———. 1969 [1953]. *Vie et moeurs des abeilles.* Trans. André Dalcq. Paris: Éditions J'ai lu.

———. 1987 [1957]. *Le Professeur des abeilles: Mémoires d'un biologiste.* Trans. Michel Martin and Jean-Paul Guiot. Paris: Belin.

Gachelin, Gabriel. 2011. "Être médecin et amateur sous les Tropiques." *Alliage,* special issue "Amateurs" no. 69: 48–61.

———. 1968. "Laveran Alphonse (1845–1922)." In *Encyclopedia Universalis.*

———. 1968. "Paludisme: découverte du parasite." In *Encyclopedia Universalis.*

Galileo. 1914 [1638]. *Dialogues Concerning Two New Sciences.* Trans. Henry Crew and Alfonso de Salvio. New York. Macmillan.

Galperin, Charles. 2006. "À l'école de la Drosophile: L'emboîtement des modèles." In Gabriel Gachelin, ed., *Les Organismes modèles dans la recherche médicale,* pp. 209–28. Paris: PUF.

Gaudry, Emmanuel. 2010. "L'entomologie légale: Une machine à remonter le temps." *Les Amis du Muséum national d' histoire naturelle,* no. 243: 36–39.

Gayon, Jean. 1992. *Darwin et l'après-Darwin, une histoire du concept de sélection naturelle.* Paris: Kimé.

———. 2006. "Les organismes modèles en biologie et en médecine." In Gabriel Gachelin ed., *Les Organismes modèles dans la recherche médicale,* pp. 9–43. Paris: PUF.

Geoffroy Saint-Hilaire, Étienne. 1796. "Mémoire sur les rapports naturels des Makis Lémur L. et description d'une espèce nouvelle de Mammifères." *Magasin encyclopédique* 1:20–50.

———. 1818. *Philosophie anatomique.* Paris: Pichon and Didier.

Ghosh, Amitav. 2008 [1996]. *The Calcutta Chromosome: A Novel of Fevers, Delirium, and Discovery.* London: Penguin.

Gillispie, Charles C. 1997. "De l'histoire naturelle à la biologie: Relations entre les programmes de recherche de Cuvier, Lamarck et Geoffroy Saint-Hilaire." In Claude Blanckaert, Claudine Cohen, Pietro Corsi, and Jean-Louis Fisher, eds., *Le Muséum au premier siècle de son histoire.* Paris: MNHN.

Goetz, Benoît. 2007. "L'Araignée, le Lézard et la Tique: Deleuze et Heidegger lecteurs de Uexküll." *Le Portique* 20, https://journals.openedition.org/leportique/1364.

Golding, William. 1954. *The Lord of the Flies.* London: Faber and Faber.

Gomel, Luc. 2003. "Jean-Henri Fabre et les fourmis." in Yves Delange, ed., *Jean-Henri Fabre, un autre regard sur l'insecte*, pp. 251–63. Rodez: Conseil général de l'Aveyron.

Gorceix, Paul. 2005. *Maurice Maeterlinck, l'arpenteur de l'invisible*. Brussels: Le Cri.

Gordon, Deborah M. 1992. "Wittgenstein and Ant-Watching." *Biology and Philosophy* 7:13–25.

——. 1996. "The Organization of Work in Social Insect Colonies." *Nature* 380:121–24.

——. 2007. "Control Without Hierarchy." *Nature* 446:143.

Gouhier, Henri. 1983. *Rousseau et Voltaire: Portraits dans deux miroirs*. Paris: Vrin.

Gouillard, Jean. 2004. *Histoire des entomologistes français (1750–1950)*. Paris: Société nouvelle des éditions Boubée.

Gould, James L., and Carol Grant Gould. 1988. *Honey Bee*. Scientific American Library. New York: Freeman.

Gould, Stephen Jay. 1989. *Wonderful Life: The Burgess Shale and the Nature of History*. New York: Norton.

——. 2002. *I Have Landed: The End of a Beginning in Natural History*. Cambridge, MA: Harvard University Press.

Gourmont, Remy de. 1903. *La Physique de l'amour*. Paris: Mercure de France.

——. 1925–1931 [1907]. "Le sadisme." In *Promenades philosophiques,* 2:269–75. Paris: Mercure de France.

Grassé, Pierre-Paul. 1959. "La reconstruction du nid et les coordinations interindividuelles chez *Bellicositermes natalensi et Cubitermes sp.*: La théorie de la stigmergie. Essai d'interprétation des termites constructeurs." *Insectes sociaux* 6, no. 1: 41–83.

——. 1985. *Zoologie*, 2d ed. 2 vols. Paris: Masson.

——, ed. 1962. *La vie et l'oeuvre de Réaumur (1683–1757)*. Paris, PUF.

Grimaldi, David. 2001. "Insect Evolutionary History from Handlisch to Hennig and Beyond." *Journal of Paleontology* 75, no. 6: 1152–60.

Guillaume, Marie. 2001. "Dis pourquoi les mouches peuvent-elles marcher au plafond?" *Insectes*, no. 122: 37.

Gusdorf, Georges. 1985. *Le savoir romantique de la Nature*. Paris: Payot.

Guyon, Étienne, ed. 2006. *L'école normale de l'an III*. Paris: Éditions rue d'Ulm.

Haldane, John Burdon Sanderson. 1949. *What Is Life?* London: Lindsay Drummond.

———. 1985 [1927]. "On Being the Right Size." In *On Being the Right Size and Other Essays.* New York: Oxford University Press.

Hamilton, W. D. 1964. "The Genetical Evolution of Social Behavior." *Journal of Theoretical Biology,* no. 7: 1–16 (part. 1), 17–52 (part 2).

Harding, Wendy, Marie Bouchet, and Laurence Talairach, eds. 2012. *Insects and Texts: Spinning Webs of Wonder.* Explora International Conference, 4–5 May 2010. Toulouse Natural History Museum/CAS (UTM).

Haüy, René-Just. 1792. "Sur les rapports de figure qui existent entre l'alvéole des abeilles et le grenat dodécaèdre." *Journal d'Histoire naturelle* 2:47–53.

Heidegger, Martin. 1995 [1983]. *The Fundamental Concepts of Metaphysics: World, Finitude, Solitude.* Trans. William McNeill and Nicholas Walker. Bloomington: Indiana University Press.

Hennig, Willi. 1987 [1965]. "Phylogenetic Systematics." *Annual Review of Entomology* 10:97–116.

Hölldobler, Bert, and Edward O. Wilson. 1996 [1994]. *Journey to the Ants.* Cambridge, MA: Belknap Press of Harvard University Press.

Hoquet, Thierry, ed. 2005. *Les Fondements de la botanique.* Paris: Vuibert.

Huber, François. 1796 [1792]. *Nouvelles observations sur les abeilles.* Geneva: Barde et Manget.

Huber, Pierre. 1810. *Recherches sur les mœurs des fourmis indigenes,* pp. xvi–328. Paris: Paschoud. English edition: 1820. *The Natural History of Ants.* Trans. J. R. Johnson. London: Printed for Longman, Hurst, Rees, Orme, and Brown. https://archive.org/details/naturalhistoryofoohube.

Huyghe, Édith, and François-Bernard Huyghe. 2006. *La route de la soie ou les empires du mirage.* Paris: Payot.

Husserl, Edmund. 1966 [1931]. *Méditations cartésiennes: Introduction à la phénoménologie.* Trans. G. Peiffer and E. Levinas. Paris: Vrin. https://archive.org/details/CartesiamMeditations.

Hutchinson, George Evelyn. 1959. "Hommage to Santa Rosalia or Why Are There So Many Kinds of Animals?," *American Naturalist* 93, no. 870: 145–59.

Israel, Giorgio, and Ana Millán Gasca, ed. 2002. *The Biology of Numbers: The Correspondence of Vito Volterra on Mathematical Biology.* Berlin: Birkhäuser.

Jaisson, Pierre. 1993. *La fourmi et le sociobiologiste.* Paris: Odile Jacob.

Jansen, Sarah. 2001a. "Histoire d'un transfert de technologie: De l'étude des insectes à la mise au point du Zyklon B." *La Recherche*, no. 340: 55–59. English: 2000. "Chemical-Warfare Techniques for Insect Control: Insect 'Pests' in Germany Before and After World War I." *Endeavour*, no. 24: 28–33.

———. 2001b. "Ameisehügel, Irrenhaus and Bordell: Insektenkunde und Degenerationdiskurs bei August Forel (1848–1931). Entomologe. Psychiater und Sexualreformer." In N. Haas, R. Nägele, and H. J. Rheinberger, eds., *Kontamination*, 141–84. Eggingen: Édition Isele.

Jaussaud, Philippe, and Edouard Brygoo. *Du jardin au muséum en 516 biographies*. Paris: Muséum national d'histoire naturelle, Publications scientifiques, 2004.

Jeannel, René. 1946. *Introduction à l'entomologie*, vol. 2: *Biologie*. Paris: Boubée.

Jolivet, Paul. 1991. "Les fourmis et les plantes: un exemple de coévolution." *Insectes*, no. 83: 3–6.

Jolivet, Gilbert. 2007. "Peut-on encore lire L'Insecte de Jules Michelet?" *Insectes*, no. 147: 9–11.

Jollivet, Servanne, and Claude Romano, eds. 2009. *Heidegger en dialogue 1912–1930: Rencontres, affinités et confrontations*. Paris: Vrin.

Jourdheuil, Pierre, Pierre Grison, and Alain Fraval. 1991. "La lutte biologique un aperçu historique." *Courrier de la cellule environnement de l'INRA*, no. 15: 37–60.

Judson, Olivia. 2006 [2002]. *Manuel universel d'éducation sexuelle à l'usage de toutes les espèces*. Paris: Seuil.

Julliard, Romain. 2017. "Science participative et suivi de la biodiversité: L'expérience Vigie-Nature." *Natures Sciences Sociétés* 25:412–17.

Jünger, Ernst. 1969 [1967]. *Chasses subtiles*. Trans. Henri Plard. Paris: Christian Bourgois.

Kafka, Franz. 2014 [1915]. *The Metamorphosis: A New Translation by Susan Bernofsky*. New York: Norton. French edition: 1955 [1915]. *La Métamorphose*. Trans. Alexandre Vialatte. Paris: Gallimard.

Kaplan, Edward K. 1977. *Michelet's Poetic Vision: A Romantic Philosophy of Nature, Man, and Woman*. Amherst: University of Massachusetts Press.

Karlson, Peter, and Martin Lüscher. 1959. "'Pheromones,' a New Term for a Class of Biologically Active Substances." *Nature*, no. 183: 55–56.

King, Lawrence J. 1975. "Sprengel." In Charles C. Gillispie, ed., *Dictionary of Scientific Biography*, 12:587–91. New York: Scribner.

Kingsland, Sharon E. 1985. *Modeling Nature, Episodes in the History of Population Ecology*. Chicago: University of Chicago Press.

Kirby, William, and William Spence. 1814. *Introduction to Entomology: Elements of the Natural History of Insects, 1814–1826*. 4 vols. London: Longman, Hurst, Rees, Orme, and Brown.

Koenig, Samuel. 1740. "Lettre de M. Koenig à M. A. B. écrite de Paris à Berne le 29 novembre sur la construction des alvéoles des abeilles . . ." *Journal Helvétique*, pp. 353–63.

Kohler, Robert E. 1994. *Lords of the Fly: Drosophila Genetics and the Experimental Life*. Chicago: University of Chicago Press.

Kropotkine, Pierre. 1902. *Mutual Aid: A Factor of Evolution*. New York: McClure Philips.

——. 1979. *L'Entr'aide un facteur de l'évolution*. Paris: Éditions de l'Entr'aide.

Lacène, Antoine. 1822. *Mémoire sur les Abeilles*. Lyon: Société royale d'agriculture de Lyon.

Lacoste, Jean. 1997. *Goethe, Science et Philosophie*. Paris: PUF.

Lamarck, Jean-Baptiste. 1801. "Discours d'ouverture du cours de zoologie, donné dans le Muséum national d'histoire naturelle, l'an VIII de la République, le 21 floréal." In *Systèmes des animaux sans vertèbres*. Paris: Déterville.

——. 1994 [1809]. *Philosophie zoologique*. Paris: Flammarion. Translation: 1914. *Zoological Philosophy*. Trans. Hugh Elliot. London: Macmillan.

Lamore, Donald H. 1969. *L'Image chez J.-H. Fabre d'après "La vie des araignées," étude stylistique*. Aix-en-Provence: La Pensée universitaire.

Lamy, Michel. 1997. *Les Insectes et les hommes*. Paris: Albin Michel.

Larrère, Catherine, and Raphaël Larrère. 1997. "Le contrat domestique." *Le Courrier de l'Environnement de l'INRA*, no. 30: 5–18.

Latour, Bruno. 1984. *Les Microbes, Guerre et paix*. Paris: Métailié.

Latreille, Pierre-André. 1798. *Essai sur l'histoire des Fourmis de la France*. Brive: Bourdeaux. Reprint: 1989. Introduction by Jean-Marc Drouin. Geneva: Champion-Slatkine and Paris: Cité des Sciences.

——. 1802. *Histoire naturelle des fourmis et recueil de mémoires et d'observations sur les abeilles, les araignées, les faucheurs et autres insectes*. Paris: Théophile Barrois Père.

———. 1810. *Considérations générales sur l'ordre naturel concernant les classes des crustacés, des arachnides et des insectes.* Paris: Schoell.

La Vergata, Antonello. 1996. "Espinas, Alfred 1844–1922. In Patrick Tort, ed., *Dictionnaire du darwinisme et de l'évolution,* 1:1402–3. Paris: PUF.

Lecointre, Guillaume, and Hervé Le Guyader. 2001. *Classification phylogénétique du vivant.* Illustrations Dominique Visset. Paris: Belin. English edition: 2006. *The Tree of Life.* Trans. Karen McKoy. Cambridge, MA: Harvard University Press.

Le Goff, Jacques. 1964. *La Civilisation de l'Occident médiéval.* Paris: Arthaud.

Le Guyader, Hervé. 2000. "Le concept de plan d'organisation: quelques aspects de son histoire." *Revue d'histoire des sciences* 53, nos. 3–4: 339–79.

Lepeletier de Saint-Fargeau, Amédée. 1836. *Histoire naturelle des Insectes,* vol. 1: *Hyménoptères.* Paris: Roret.

Leraut, Patrice, and Gilles Mermet. 2003. *Regard sur les insectes.* Paris: MNHN, Imprimerie nationale.

Lesser, Friedrich Christian. 1742. *Théologie des insectes, ou Démonstration des perfections de Dieu dans tout ce qui concerne les insects.* Ed. and trans. Pierre Lyonet. La Haye: Jean Swart.

Lestel, Dominique. 1985. "Les Fourmis dans le panoptique." *Culture technique,* no. 14: 125–31.

———. 2003 [2001]. *Les Origines animales de la culture.* Paris: Flammarion.

Lévi-Strauss, Claude. 2002 [2001]. "Guillaume Lecointre et Hervé Le Guyader, Classification phylogénétique du vivant." *L'Homme,* no. 162 (April-June). http://lhomme.revues.org/index169.html.

Lhoste, Jean. 1987. *Les Entomologistes français, 1750–1950.* Guyancourt: INRA-OPIE.

Lhoste, Jean, and Janine Casevitz-Weulersse, eds. 1997. *La fourmi.* Paris: Muséum national d'histoire naturelle.

Lhoste, Jean, and Bernard Henry. 1990. "Les insectes dans l'art d'Extrême-Orient. *Insectes,* nos. 76, 77: 16–17, 16–17.

Linnaeus, Carl von. 1758. *Systema naturae,* 10th ed. Stockholm: Salvius.

———. 1758 [1744]. *Systema naturae.* In *Opera varia in quibus continentur Fundamenta Botanica, Sponsalia plantarum, Systemae Naturae, Lucae.* Reprint of the 4th edition. Leyde: Typographia Justiniana.

———. 1966 [1751]. *Philosophia Botanica.* Lehre: J. Cramer.

———. 1972. *L'équilibre de la nature*. Ed. Camille Limoges. Trans. Bernard Jasmin. Paris: Vrin.

Lotka, Alfred. 1925. *Elements of Physical Biology*. Baltimore: Williams and Wilkins. Revised edition 1956. *Elements of Mathematical Biology*. New York: Dover.

Lourenço, Wilson R. 2008. "La biologie reproductrice chez les Scorpions." *Les Amis du Muséum national d'histoire naturelle*, no. 236: 49–52.

Lubbock, John. 1882. *Ants, Bees, and Wasps: A Record of Observations on the Habits of the Social Hymenoptera*, 3d ed. London: Kegan Paul.

Lupoli, Roland. 2011. *L'insecte medicinal*. Fontenay-sous-Bois: Ancyrosoma.

Lustig, Abigail. 2004. "Ants and the Nature of Nature in August Forel, Erich Wasmann, and William Morton Wheeler." In L. Daston and F. Vidal, eds., *The Moral Authority of Nature*, pp. 282–307. Chicago: University of Chicago Press.

Maderspacher, Florian. 2007. "All the Queen's Men." *Current Biology* 17, no. 6: 191–95.

Maeterlinck, Maurice. 1919. "Le monde des insectes." In *Les sentiers dans la montagne*, pp. 81–116. Paris: Fasquelle.

———. 1927 [1926]. *La vie des termites*. Paris: Fasquelle.

———. 1930. *La vie des fourmis*. Paris: Fasquelle. Trans. Bernard Miall. https://archive.org/stream/lifeoftheantbymaо15682mbp/lifeoftheantby maо15682mbp_djvu.txt.

———. 1963 [1901]. *La vie des abeilles*. Paris: Fasquelle.

Magnin, Antoine. 1911. *Charles Nodier naturaliste: Ses oeuvres d'histoire naturelle publiées et inédites*. Paris: Libraire scientifique Hermann et fils.

Magnin-Gonze, Joëlle. 2009 [2004]. *Histoire de la Botanique*, 2d ed. Paris: Delachaux and Niestlé.

Mandal, Sandip, Ram R. Sarkar, and Sinha Somdatta. 2011. "Mathematical Models of Malaria: A Review." *Malaria Journal* 10, https://doi.org/10 .1186/1475-2875-10-202.

Mandeville, Bernard. 1990 [1714]. *La fable des abeilles, ou les vices privés font le bien public*. Ed. Paulette Carrive and Lucien Carrive. Paris: Vrin.

Marais, Eugène. 1950 [1938]. *Moeurs et coutumes des Termites: La Fourmi blanche de l'Afrique du Sud*. Trans. S. Bourgeois. Paris: Payot.

Maraldi, Giacomo Filippo. 1731 [1712]. "Observations sur les Abeilles." *Mémoires de l'Académie royale des sciences*, pp. 297–331. Paris.

Marchal, Hugues. 2007. "Le conflit des modèles dans la vulgarisation ento-mologique: L'exemple de Michelet, Flammarion and Fabre." *Romantisme*, no. 138: 61–74.

Marchenay, Philippe, and Laurence Bérard. 2007. *L'Homme, l'abeille et le miel*. Romagnat: De Borée.

Marx, Karl. 1969 [1867]. *Le Capital*, vol. 1. Trans. Joseph Roy. Paris: Flammarion.

Massis, Henri. 1924. *Jugements II: André Gide, Romain Rolland, Georges Duhamel, Julien Benda, les chapelles littéraires*. Paris: Plon.

Merleau-Ponty, Maurice. 1995. *La Nature, Notes de cours 1957–1958, Collège de France*. Ed. P. Seglard. Paris: Seuil.

Michelet, Jules. 1998. *Correspondance générale*, vol. 8: *1856–1858*. Ed. by Louis Le Guillou. Paris: Honoré Champion.

——. 2011 [1858]. *L' insecte*. Paris: Hachette. Ed. Paule Petitier. Sainte-Marguerite-sur-Mer: Édition des Équateurs. English edition: 1875. *The Insect*. Trans. W. H. Davenport Adams. London: T. Nelson.

Milgram, Maurice, and Henri Atlan. 1983. "Probabilistic Automata as a Model for Epigenesis of Cellular Networks." *Journal of Theoretical Biology*, no. 103: 523–47.

Miller, Peter. 2007. "The Genius of Swarms." *National Geographic* 212, no. 1: 126–47.

Miller, Philip. 1759. "Generation." *The Gardeners Dictionary*, 7th ed. vol. 1. London. Printed for the author.

Moggridge, Johann Treherne. 1873. *Harvesting Ants and Trap Down Spiders: Notes and Observations on Their Habits and Dwellings*. London: L. Reeve.

Montaigne, Michel de. 1962. *Essais*. 2 vols. Paris: Garnier.

Morange, Michel. 1994. *Histoire de la biologie moléculaire*. Paris: La Découverte.

Morgan, Thomas Hunt. 1965. "The Relation of Genetics to Physiology and Medicine" [1933]. In *Nobel Lecture, Physiology or Medicine: 1922–1941*, pp. 313–28. Amsterdam, Elsevier.

Morgan, Thomas Hunt, Alfred Sturtevant, Hermann Joseph Muller, and Calvin Bridges. 1915. *The Mechanism of Mendelian Heredity*. New York: Henry Holt.

Mornet, Daniel. 1911. *Les Sciences de la Nature en France au 18ᵉ siècle, un chapitre de l'histoire des idées*. Paris: Armand Colin.

Mulsant, Étienne. 1830. *Lettres à Julie sur l'entomologie*. 2 vols. Lyon: Babeuf.

Népote-Desmarres, Fanny. 1999. *La Fontaine: Fables*. Paris: PUF.

Nodier, Charles. 1982 [1832]. *La Fée aux miettes* [1832]. *Smarra* [1821]. *Trilby* [1822]. Paris: Gallimard.

Nuridsany, Claude, and Marie Pérennou. 1996. *Microcosmos, le peuple de l'herbe*. Paris: La Martinière.

Orr, Linda. 1976. *Jules Michelet, Nature, History, and Language*. Ithaca: Cornell University Press.

Ostachuk, Agustín. 2013. "El Umwelt de Uexküll y Merleau-Ponty." *Ludus Vitalis* 21, no. 39: 45–65.

Pain, Janine. 1988. "Les phéromones d'Insectes: 30 ans de recherché." *Insectes*, no. 69: 2–4.

Pappus d'Alexandrie. 1982 [1932]. *La Collection Mathématique*. Trans. Paul ver Eecke. Paris: A. Blanchard.

Pascal, Blaise. 1954. *Pensées*. In *Oeuvres completes*. Ed. Jacques Chevalier. Paris: Gallimard.

Passera, Luc. 1984. *L'organisation sociale des Fourmis*. Toulouse: Éditions Privat.

——. 2006. *La véritable histoire des Fourmis*. Paris: Fayard.

Peckham, George W., and Elizabeth G. Peckham. 1905. *Wasps, Solitary and Social*. Boston: Houghton, Mifflin.

Pelozuelo, Laurent. 2007. "Mushi." *Insectes*, no. 145: 9–12.

——. 2008. "La Femme des sables: Regards d'entomologistes." *Inf'opie-mp*, no. 8. http://www.insectes.org/opie/pdf/685_pagesdynadocs49639ceac 4954.pdf.

Perrin, Hélène. 2008. "Hymnes au charançon." *Insectes*, no. 148, p. 11–13.

——. 2009. "Coton, charançon, chansons . . ." *Mémoires de la S.E.F.*, no. 8: 67–69.

——. 2010. "Des charançons à la rescousse." *Insectes*, no. 159: 23–27.

Perron, Jean-Marie. 2006. "Connaissez-vous les *Lettres à Julie? Antennae: Bulletin de la Société d'entomologie du Québec* 13, no. 1: 5–7. http://www.seq .qc.ca/antennae/archives/articles/Article_13-1-Lettres_a_Julie.pdf.

Perru, Olivier. 2003. "La problématique des insectes sociaux: Ses origines au 18ᵉ siècle et l'oeuvre de Pierre-André Latreille." *Bulletin d'histoire et d'épistémologie des sciences de la vie* 10, no. 1: 9–38.

Petit, Annie. 1988. "La diffusion des sciences comme souci philosophique: Bergson." In B. Bensaude-Vincent and C. Blondel, eds., *Vulgariser les sciences (1919–1939). Acteurs, projets, enjeux. Cahiers d'histoire et de philosophie des sciences*, no. 24: 15–32.

———. 1991. "La philosophie bergsonienne, aide ou entrave pour la pensée biologique contemporaine." *Uroboros: Revista international de filosofía de la biología* 1, no. 2: 177–79.

———. 1999. "Animalité et humanité: Proximité et altérité selon H. Bergson." *Revue européenne des sciences sociales* 37, no. 115: 171–83.

———. 2007. "Science et synthèse selon Marcelin Berthelot." In Jean-Claude Pont, Laurent Freland, Flavia Padovani, and Lilia Slavinskaia, eds., *Pour comprendre le 19ᵉ siècle: Histoire et philosophie des sciences à la fin du siècle,* pp. 3–42. Florence: Leo Olschki.

Picq, Pascal. 2003. "Le réel des animaux." In G. Cohen-Tannoudji and E. Noël, eds., *Le réel et ses dimensions,* 109–27. Les Ulis: EDP Sciences.

Pieron, Julien. 2010. "Monadologie et/ou constructivisme: Heidegger, Deleuze, Uexküll. *Bulletin d'analyse phénoménologique* 6, no. 2: 86–117. http://popups.ulg.ac.be/bap/document.php?id=384.

Pilet, P. E. 1972. "Forel Auguste Henri . . ." In Charles C. Gillispie, ed., *Dictionary of Scientific Biography,* 5:73–74. New York: Scribner.

Pinault-Sørensen, Madeleine. 1991. *Le Peintre et l'histoire naturelle.* Paris: Flammarion.

Plato. 1979. *Plato's Apology of Socrates: An Interpretation, with a New Translation.* Ed. and trans. Thomas G. West. Ithaca, NY: Cornell University Press.

Pliny the Elder. 1848–1850. *Histoire naturelle,* book 11. Trans. French Émile Littré. Paris: Dubochet. French online version edited by Philippe Remacle with Agnès Vinas in Méditerranées: http://remacle.org/bloodwolf /erudits/plineancien/. English edition: http://www.perseus.tufts.edu /hopper/text?doc=Perseus%3Atext%3A1999.02.0137%3Abook%3D11%3 Achapter%3D4.

Pluche, Noël Antoine. 1732. *Le Spectacle de la Nature ou Entretiens sur les particularités de l'Histoire naturelle qui ont paru les plus propres à rendre les jeunes gens curieux et à leur former l'esprit.* Paris: V. Estienne.

Poincaré, Henri. 1908. *Science et méthode.* Paris: Flammarion. English edition: 1914. *Science and Method.* Trans. Francis Maitland. London: Thomas Nelson.

Poliakov, Léon. 1968. *Histoire de l'antisémitisme: De Voltaire à Wagner.* Paris: Calmann-Lévy.

Poupart, François. 1745 [1704]. "Histoire du Formica-leo." In *Mémoires de l'Académie royale des sciences,* pp. 215–46. Paris: Gabriel Martin, Jean-Baptiste Coignard, and Hippolyte-Louis Guérin.

Prete, Frederick R. 1990. "The Conundrum of the Honey Bees: One Impediment to the Publication of Darwin's Theory." *Journal of the History of Biology* 23, no. 2: 271–90.

———. 1991. "Can Females Rule the Hive? The Controversy Over Honey Bee Gender Roles in British Beekeeping Texts of the Sixteenth-Eighteenth Centuries." *Journal of the History of Biology* 24, no. 1: 113–44.

Proust, Marcel. 1954 [1913]. *Du côté de chez Swann,* vol. 1 of *À la recherche du temps perdu.* Paris: Gallimard.

———. 1954 [1921]. *Sodome et Gomorrhe,* vol. 2 of *À la recherche du temps perdu.* Paris: Gallimard.

Punnett, Reginald. 1915. *Mimicry in Butterflies.* Cambridge, Cambridge University Press.

Quatrefages, Armand de. 1854. *Souvenirs d'un naturaliste,* vol. 2. Paris: Masson.

Radelet de Grave, Patricia. 1998. "La moindre action comme lien entre la philosophie naturelle et la mécanique analytique. Continuité d'un questionnement." *LLULL* 21:439–84.

Rameaux, Jean-François. 1858. "Des lois suivant lesquelles les dimensions du corps dans certaines classes d'animaux déterminent la capacité et les mouvements fonctionnels des poumons et du cœur." *Mémoires couronnés et mémoires des savants étrangers publiés par l'Académie royale de Belgique* 29:3.

Rameaux, Jean-François, and Frédéric Sarrus. 1838–1839. "Rapport sur un mémoire adressé à l'Académie royale de médecine par MM. Sarrus, professeur de mathématiques à la faculté des sciences de Strasbourg, et Rameaux, docteur en médecine et ès sciences." *Bulletin de l'Académie royale de médecine,* 3:1094–100.

Ratcliff, Marc. 1996. "Naturalisme méthodologique et science des mœurs animales au 18ᵉ siècle." *Bulletin d'Histoire et d'Épistémologie des sciences de la vie* 3, no. 1: 17–29.

Raulin-Cerceau, Florence, with Bénédicte Bilodeau. 2009. *Les Origines de la vie: Histoire des idées.* Paris: Ellipses.

Ray, John. 1977 [1717]. *The Wisdom of God Manifested in the Works of the Creation.* Ed. R. Harbin, for William Innys. New York: Arno.

Réaumur, René Antoine Ferchault. 1734–1742. *Mémoires pour servir à l'histoire des insectes*. 6 vols. Paris: Imprimerie royale.

———. 1926. *The Natural History of Ants: From an Unpublished Manuscript in the Archives of the Academy of Sciences of Paris;* French text pp. 41–128, English trans. pp. 129–217. Ed. and trans. William Morton Wheeler. New York: Knopf. Facsimile reprint 1977. New York: Arno.

———. 1928/1929. "Histoire des fourmis." In *Mémoires pour servir à l'histoire des insectes,* vol. 7. Paris: Paul Lechevalier.

Revel, E. 1951. *J.-H. Fabre: L'Homère des Insectes*. Paris: Delagrave.

Rigol, Loïc. 2005. "Alphonse Toussenel et l'éclair analogique de la science des races." *Romantisme* 4, no. 130: 39–53.

Robert, Paul-André. 2001 [1936]. *Les insectes*. Ed. J. d'Aguilar. Lausanne: Delachaux et Niestlé.

Robillard, Tony, and Laure Desutter. 2008. "Clarification of the Taxonomy of Extant Crickets of the Subfamily Eneopterinae (Orthoptera: Grylloidea; Gryllidae)." *Zootaxa* 1789:66–68. http://www.researchgate.net/publication/228507600_Clarification_of_the_taxonomy_of_extant_crickets_of_the_subfamily_Eneopterinae_(Orthoptera_Gryl-loidea_Gryllidae).

Rollard, Christine, and Vincent Tardieu. 2011. *Arachna: Les voyages d'une femme araignée*. Paris: MNHN/Belin.

Roman, Myriam. 2007. "Histoire naturelle et représentation sociale après 1848 (Toussenel/Michelet)." Second study day dedicated to the animal in the nineteenth century. Ed. Paule Petitier. University Paris VII–Denis Diderot. http://groupugo.div.jussieu.fr.

Ross, Ronald. 1902. *Mosquito Brigades and How to Organize Them*. New York: Longmans, Green.

Roughgarden, Joan. 2012. *Le gène généreux: Pour un darwinisme coopératif.* Trans. Thierry Hoquet. Paris: Seuil. Original: 2009. *The Genial Gene: Deconstructing Darwinian Selfishness*. Berkeley: University of California Press.

Rousseau, Jean-Jacques. 1921. *Emile, or Education*. Trans. Barbara Foxley, M.A. New York: Dutton.

———. 1969 [1762]. *Émile, ou De l'éducation*. In *Oeuvres complètes,* vol. 4. Paris: Gallimard.

Ruelland, Jacques. 2004. *L'empire des gènes*. Paris: ENS Éditions.

Rüting, Torsten. 2004. "History and Significance of Jacob von Uexküll and His Institute in Hamburg." *Sign Systems Studies* 32, nos. 1, 2: 35–72.

Sartori, Michel, and Daniel Cherix. 1983. "Histoire de l'étude des Insectes Sociaux en Suisse à travers l'oeuvre d'Auguste Forel." *Bulletin de la Société Entomologique de France*, 150th anniversary, 88:66–74.

Schlanger, Judith. 1971. *Les métaphores de l'organisme*. Paris: Vrin.

Schmidt-Nielsen, Knut. 1984. *Scaling: Why Animal Size Is So Important*. Cambridge: Cambridge University Press.

Schmitt, Stéphane. 2004. *Histoire d'une question anatomique: Le problème des parties répétées*. Paris: Publications scientifiques du MNHN.

Schuhl, Pierre Maxime. 1947. "Le thème de Gulliver et le postulat de Laplace." *Journal de psychologie* 40, no. 2: 169–84.

Secord, Jim, ed. *Darwin Correspondence Project*. Cambridge University. https://www.darwinproject.ac.uk.

Séméria, Yves. 1985. "Le philosophe et l'insecte. Nicolas Malebranche, 1638–1715: Ou l'entomologiste de Dieu." *Supplément du Bulletin mensuel de la Société linnéenne de Lyon*, no. 1: i–vi.

Serres, Michel. 1995 [1992]. *The Natural Contract*. Trans. Elizabeth MacArthur and William Paulson. Ann Arbor: University of Michigan Press.

Serres, Olivier de. 2001 [1600]. *Le théâtre d'agriculture et mésnage des champs*. Le Méjan: Actes Sud.

Siganos, André. 1985. *Les mythologies de l'insecte: Histoire d'une fascination*. Paris: Librairie des Méridiens.

Sigrist, René, Vincent Barras, and Marc Ratcliff. 1999. *Louis Jurine, chirurgien et naturaliste (1751–1819)*. Geneva: Georg.

Sleigh, Charlotte. 2001. "Empire of the Ants: H. G. Wells and Tropical Entomology." *Science as Culture* 10, no. 1: 33–71.

——. 2003. *Ant*. London: Reaktion.

Smeathman, Henry. 1786. *Mémoire pour servir à l'histoire de quelques insectes connus sous les noms de termes ou fourmis blanches*. Paris. Golden edition 1781. *Philosophical Transactions, Royal Society*, vol. 21.

Smith, Adam. 2009 [1776]. *Recherches sur la nature et les causes de la richesse des nations*. Ed. Jacques Valier. Trans. Germain Garnier. Paris: Le Monde/Flammarion.

Smith, David L., Katherine E. Battle, Simon I. Hay, Christopher M. Barker, Thomas W. Scott, and F. Ellis McKenzie. 2012. "Ross, MacDonald and a Theory for the Dynamics and Control of Mosquito-Transmitted Pathogens." *PLOS Pathog* 8, no. 4, https://doi.org/10.1371/journal.ppat.1002588.

Smith, Ray, Thomas Mittler, and Carroll Smith. 1973. *History of Entomology*. Palo Alto: Entomological Society of America Annual Review.

Sprengel, Christian Konrad. 1793. *Das entdeckte Geheimnis der Natur im Bau und in der Befruchtung der Blumen*. Berlin: F. Vieweg.

Stafleu, Frans A. 1971. *Linnaeus and the Linnaeans. The Spreading of Their Ideas in Systematic Botany: 1735–1789*. Utrecht: Oostoek.

Swammerdam, Jan. 1758. *Histoire naturelle des insectes, traduite du Biblia naturae avec 36 planches et des notes*. Dijon: Desventes.

Swammerdam, Jan et al. 1792. *The Natural History of Insects*. Perth. Printed by R. Morison.

Swift, Jonathan. 1954 [1726/1735]. *Gulliver's Travels*. London: J. M. Dent.

Tassy, Pascal. 1991. *L'Arbre à remonter le temps*. Paris: Christian Bourgois.

——. 2000. *Le Paléontologue et l'Évolution*. Paris: Le Pommier.

Théophraste. 2003. *Recherches sur les plantes*, vol. 1. Trans. Suzanne Amigues. Paris: Les Belles Lettres.

Theraulaz, Guy, and Éric Bonabeau. 1999. "A Brief History of Stimergy." *Artificial Life* 5:97–116.

Theraulaz, Guy, Éric Bonabeau, and Jean-Louis Deneubourg. 1998. "Les Insectes architectes ont-ils leur nid dans la tête?" *La Recherche*, no. 313: 84–90.

Thibaud, Jean-Marc. 2010. "Les Collemboles, ces Hexapodes vieux de 400 millions d'années, cousins des Insectes, si communs, mais si méconnus." *Les Amis du Muséum national d'histoire naturelle*, no. 242: 20–23.

Thompson, D'Arcy W. 1992 [1917/1961]. *On Growth and Form*. Preface by Stephen Jay Gould. Cambridge: Cambridge University Press.

Thorpe, Vanessa. 2012. "Richard Dawkins in Furious Row with E. O. Wilson." *Observer*, 24 June.

Tinbergen, Nikolaas. *La Vie sociale des animaux*. Paris: Payot, 1967.

Torlais, Jean. 1961. *Un esprit encyclopédique en dehors de l'Encyclopédie: Réaumur*. Paris: Albert Blanchard.

Tort, Patrick. 1996. "Forel Auguste Henri 1848–1931." In Patrick Tort, ed., *Dictionnaire du Darwinisme et de l'évolution*, 2:1705–10. Paris: PUF.

——. 2002. *Fabre: Le miroir aux insectes*. Paris: Vuibert/ADAPT.

Toussenel, Alphonse. 1859. *L'esprit des bêtes: Le monde des oiseaux, ornithologie passionnelle*. Paris: Librairie phalanstérienne.

Trembley, Jacques, ed. 1987. *Les savants genevois dans l'Europe intellectuelle du XVIIe au milieu du 19e siècle*. Geneva: Éditions du Journal de Genève.

Uexküll, Jacob von. 1965 [1934]. *Mondes animaux et mondes humains, suivi de théorie de la signification*. Trans. Philippe Muller. Paris: Gonthier.

———. 2010. *A Foray Into the Worlds of Animals and Humans*. Trans. Joseph D. O'Neil. Minneapolis: University of Minnesota Press.

Utamaro, Kitagawa. 2009 [1788]. *Album d'Insectes choisis: Concours de poèmes burlesques des myriades d'oiseaux*. Trans. Christophe Marquet. Arles: Éditions Philippe Picquier/INHA.

Valk, Vincent. 2007. "Albert Einstein Ecologist?" *Gelf Magazin*, 25 April. http://www.gelfmagazine.com/archives/albert_einstein_ ecologist.php.

Vanden Eeckhoudt, Jean-Pierre. 1965. *Visages d'insectes*. Paris: L'école des loisirs.

Vandermeulen, David. 2005, 2007, 2010, 2014. *Fritz Haber*. 4 vols. Tournai: Delcourt.

Veuille, Michel. 1997 [1986]. *La Sociobiologie*, 2d ed. Paris: PUF.

Villemant, Claire. 2005. "Les nids d'abeilles solitaires et sociales." *Insectes*, no. 137: 13–17.

Virey, Julien-Joseph. 1819. "Société des animaux." In *Nouveau dictionnaire d'histoire naturelle*, 31:358–59. Paris: Déterville.

Virgil. 1994. *Géorgiques*. Ed. R. Lessueur. Trans. E. Saint-Denis. Paris: Les Belles Lettres.

Voltaire. 1960 [1752]. "Micromégas." In *Romans et contes*, pp. 96–113. Paris: Garnier.

Volterra, Vito, and Umberto D'Ancona. 1935. *Les Associations biologiques au point de vue mathématique*. Paris: Hermann.

Wallace, Alfred Russell. 1897 [1889]. *Darwinism, an Exposition of the Theory of Natural Selection with Some of Its Applications*. London: Macmillan.

Wells, Herbert George. 1977 [1905]. *The Empire of the Ants and Other Short Stories*. New York: Scholastic.

Werber, Bernard. 1991. *Les fourmis*. Paris: Albin Michel.

Wheeler, William Morton. 1926. *Les Sociétés d'insectes: Leur origine. Leur évolution*. Paris: Doin.

———. 1928. *The Social Insects: Their Origin and Evolution*. London: K. Paul, Trench, and Trubner.

Wilson, Edward O. 1975. *Sociobiology: The New Synthesis*. Cambridge, MA: Harvard University Press.

——. 1976. "The Central Problem of Sociobiology." In R. May, ed., *Theoretical Ecology: Principles and Applications,* pp. 205–17. Oxford: Blackwell.

——. 1978. "Introduction: What is Sociobiology?" In Michael Gregory, Anita Silvers, and Diane Sutch, eds., *Sociobiology and Human Nature: An Interdisciplinary Critique and Defense,* pp. 1–12. San Francisco: Jossey Bass.

——. 1984. "Clockwork Lives of the Amazonian Leaf-cutter Army." *Smithsonian* 15, no. 7: 92–100.

Winsor, Mary P. 1976. "The Development of Linnaeus Insect Classification." *Taxon* 25, no. 1: 57–67.

Xenophon. 2008 [1949]. *Économique.* Trans. Pierre Chantraine. Paris: Les Belles Lettres.

Yavetz, Ido. 1988. "Jean-Henri Fabre and Evolution: Indifference or Blind Hatred?" *History and Philosophy of the Life Sciences* 10:3–36.

——. 1991. "Theory and Reality in the Work of Jean-Henri Fabre." *History and Philosophy of the Life Sciences* 8:33–72.

INDEX

CPSIA information can be obtained
at www.ICGtesting.com
Printed in the USA
LVHW041131111219
640145LV00004B/10/P